How I Made the World:

Shaping a View of Landscape

THE
UNIVERSITY
OF HULL
PRESS

Cover illustration: Cycladic invention in Tinachrome.
Painting in oils by Jay Appleton, reproduced by permission of the
owner, Tina Harrison (see also Fig. 36).

FRONTISPIECE. Jay and Iris Appleton in Cottingham, 1990.
Photo by Charles G. Appleton.

How I Made the World:

Shaping a View of Landscape

Jay Appleton

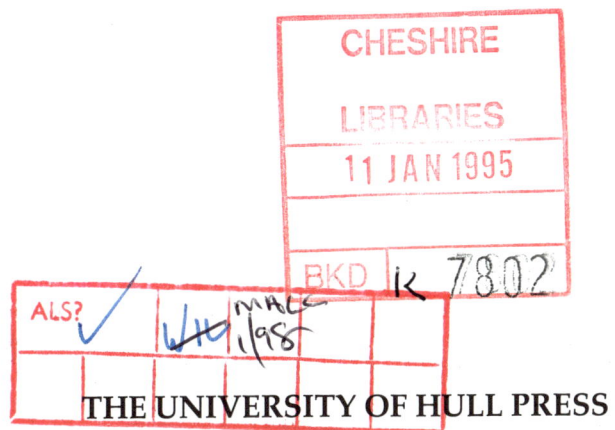
THE UNIVERSITY OF HULL PRESS

© Jay Appleton

British Library Cataloguing in Publication Data

A catalogue record for this book is available from the British Library.

ISBN 0 85958 620 0
Published 1994

Phototypeset in 11 on 12pt Times by Gem DTP, 37 Hunter Road, Elloughton, HU15 1LG and printed by the Central Print Unit, the University of Hull.

On this, my Golden Wedding Anniversary, I dedicate this book to my family in recognition of the contributions they have made, wittingly or unwittingly, to the making of my world.

Cottingham
East Yorkshire.

5 August 1993.

CONTENTS

PREFACE

'In the beginning God created the heaven and the earth.
And the earth was without form, and void.'

Genesis 1:1.

It's a safe bet that our successors will argue about the origin of the world as vehemently as our predecessors or, for that matter, ourselves. Some may indeed question whether there is a 'real' world at all, a world existing independently of our own observations and conceptions, but that's a question which we shall leave to the philosophers. For our purposes I shall assume the reality of such a world, however it came into existence; but there can be no argument about the fact that all children, as soon as they are born, must begin to build up their own picture of it. That picture will change and develop as they go through life. Every experience of their environment will affect it, confirming it to be true, exposing it as false, or, more often, suggesting the need for some fine-tuning to bring their own created image more closely into line with what we call reality.

In so far as our behaviour is conditioned by our surroundings the model is more important than the original. For all practical purposes the picture is our world. It's on the picture and not the reality that we base all our decisions. The picture may be, and usually is, wrong, and that's why we need to be perpetually adjusting it by environmental perception. While most of us are content to accept this as so obvious as not to require justification, psychologists and other behavioural scientists have studied more closely the processes by which such adjustments take place. There is a large and growing literature on the subject, but it concerns itself principally with establishing general principles from the analysis of a very large number of individual cases, every one of which, while it may share common characteristics with others, will be, as an individual case, unique.

The study of either the general principles or the individual case is incomplete by itself. General principles alone don't give us an authentic picture of anybody who actually exists any more than one individual can accurately represent the rest of humanity. We need to look at both the broad picture of what we may call 'usual behaviour' and that unique association of habits which is only to be found in a real person; and because the literature on environmental perception and adaptation is massively dominated by the former, the operation of general principles, there is an urgent need for more studies of the latter kind, individual case-studies, to restore the balance.

As well as differing in our habits of perceiving the landscape around us we differ also in our preferences for particular kinds of landscape. Perception and preference are closely related, and in the following pages we shall be looking at both as they have developed side by side in one person. If I were a practising psychologist or psychiatrist maybe I would go carefully through my files and my casebooks and choose as my subject some patient about whom I had amassed a voluminous record from which to draw the raw material of my study. Being, however, only an interested amateur, a geographer with a nosy interest in other people's disciplines, I'm forced to choose the one individual of whom I have the most complete record of information, that is to say myself.

The task I have set myself isn't an easy one and for some time I was hesitant to embark on it. I was, however, encouraged by the example of one or two other geographers who have preceded me. The use of an autobiographical approach in a serious study of environmental perception can be found in a book by John Eyles of Queen Mary College, London, entitled *Senses of Place*. He describes it as 'an investigation of the forms that sense of place may take and an attempt to relate these forms to the totality of individual lives' (Eyles, 1985, 6), and he uses an autobiographical approach to get into his subject.

Eyles and I have quite a lot in common and, although I didn't know it until I read his book, we have both taught in the same department of the same Australian university. But there are two main differences between his book and mine. First, his is a more conventionally academic work, (based on a Ph.D. thesis in fact), fully referenced and aimed at the serious student of that branch of geography which now goes under the name of 'humanistic', whereas mine is less specifically aimed at the discipline of geography. Secondly, Eyles is very much concerned with the social rather than the visual environment which is what I shall be dealing with in the following pages.

Perhaps the principal source from which I've drawn confidence is one of my former students, Jacquelin Burgess of University College, London, who some time ago convinced me that, if we're serious in our intention to find out how people view their environment, it isn't good enough to dismiss potential evidence simply on the grounds that it's 'non-academic' in form and appearance. She and her collaborators (Harrison, Limb and Burgess, 1986) have adapted the small-group techniques, successfully used in psychotherapy, to the study of attitudes of different kinds of people to different kinds of landscape, compiling many hours of tape-recordings of 'chat', often ungrammatical and casually communicated, sometimes ill-expressed and not always directed to the point under discussion, in short, irrelevant. As orthodox serious literature most of this material would be a non-starter, but as a verbal bran-tub containing hidden treasure for those with the initiative to explore it and the expertise to recognise what they find, it's a rich repository of information which can throw quite a new light on the subject. All this encourages me not to be afraid to kick over the traces of academic convention.

Most people, when they pick up a book, do so with certain expectations of what it's going to be about, and one of the most common sources of disappointment is the discovery that it's about something else. Let me therefore tell you what it *isn't* about and maybe save you wasting your time by reading any further. First, it doesn't make any pretence to the sort of authority that you would expect from a highly scientific treatise written by a specialist. It isn't the definitive overview of a recognised field of knowledge. It certainly isn't an academic textbook. It won't use a lot of abstruse technical terminology and it won't use more abstract terms than are necessary. Rather it will be about practical things like going to places, looking at them, wondering about them, forming likes and dislikes about them, and so on. Towards the end we shall be looking at more theoretical ideas, but it won't assume more than a minimum level of technical knowledge on the part of the reader, and if there are unusual technical terms they will be explained. In other words it's not a book to be read only by specialists.

Secondly it's not an autobiography in the generally accepted sense of the word. It doesn't set out to tell comprehensively the story of my life. People, places and ideas come into it in so far as they have a bearing on the development of my habits of perception and the formation of my tastes in landscape, so the criterion of what goes in and what stays out is very different from that which would generally guide the writer of an autobiography.

Thirdly, you mustn't think of it as yet another of those studies of East Anglian villages in the olden days, of which a proliferation has appeared during recent years. It isn't a Norfolk man's answer to *Akenfield* (Blythe, 1969), though a Norfolk village does play a central role in Chapter 1 and keeps making sporadic re-appearances. This is because it formed the stage on which much of the plot was set at a crucial period, or, to change the metaphor, the crucible in which my early habits and tastes were formed. Even the parts which are concerned with the village are in no sense a sociological survey; it's more of a micro-geography of what my sister and I came to regard as our own territorial patch. It was the first and most important source of raw material from which I created my own world.

In the beginning, then, *I* created the heaven and the earth. When I emerged from my mother's womb my own world, the one inside my little brain, was indeed without form and void. Although I can remember nothing of those early days, I now know that I must from the beginning have been exposed to many sensations, perceived through all the senses, from which I needed to begin making order out of chaos. My total dependence on others, chiefly my mother, was to be gradually broken down and replaced by an increasing dependence on this developing picture of the world in which I had to survive. My earliest recollections go back to a stage at which the broad outlines of this picture were already formed, and had been for some time. But I can remember enough of my childhood to be able to understand, with hindsight, quite a lot about the process by which that picture underwent change after change and left me a unique individual, conforming to the general pattern of environmental adaptation characteristic of all members of my species but pursing distinctive variations on that general theme which are peculiar to myself.

In particular I think I can now understand some of the reasons why I have acquired a sort of portfolio of tastes and preferences in landscape, all of which are probably shared by some other individuals but not in that particular mix which is uniquely my own. The story of how I acquired this mix is the subject of this book, but I hope that, as you read it, it may also shed some light on how *you* put together the collection of habits, tastes and preferences which is uniquely *yours*. Think of it like an algebraic equation in which you can substitute for the symbols your own set of values, different from mine, without upsetting the validity of the formula.

The structure of the book is quite simple. Part I is a roughly chronological though highly selective account of my life up to the age of about thirty. It's the part which most closely resembles a straightforward autobiography. Perhaps I should warn you that, when you encounter the events and processes which it describes, you may not immediately see their relevance to the subject of the book. Some of them may have crept in on dubious credentials where a more rigid interpretation of my own guidelines should have excluded them, but many of them are taken up again later when we have developed a broader framework into which they can be fitted.

In Part II the chronological approach is largely abandoned and we shall look at certain selected topics which have assumed importance in that model of the world which I had constructed by the time I had settled down to embark on my chosen career. Part III takes up the story at the point where I was beginning to formulate explanations of the phenomena described in the first two Parts, asking, first, what sort of a mind I had brought to undertake the task (Chapter 8), secondly, what sort of explanatory theories emerged (Chapter 9) and, lastly, (Chapter 10) how these fitted into the wider pattern of research being pursued in environmental aesthetics and how they affected my own reactions to landscapes perceived for the first time.

One problem arises from this structure and that is the difficulty of maintaining anything like a consistent style. The sort of language normally used for discussing theoretical questions is not the sort one would choose for an autobiography, and since the emphasis does change as the book progresses, I am aware that I haven't solved this to my own satisfaction, and therefore probably not to yours. Please remember that, since a major part of my argument is that all observation of, and therefore all writing about, landscape is highly subjective, it would be pointlessly misleading to use those devices which may be legitimately employed elsewhere but would here serve only to clothe subjective statements in a thin disguise of spurious objectivity, such as choosing the third person when one is referring to the first. 'When the author was in Timbuktu' actually means 'When I was in Timbuktu', and 'in the opinion of the present writer' actually means 'I think'. Similarly, who is 'the reader' if it isn't you?

ACKNOWLEDGEMENTS

I am conscious of having drawn heavily on the help of numerous friends and relations who may or may not be aware of the contributions they have made to this book. Those who are mentioned here have performed some particular service for which I wish to record my gratitude.

Of the three people who chiefly influenced my habits of perception in the early days, my mother and father passed away in 1941 and 1954 respectively; my sister, Helen, now Mrs W. G. Cook, is once again resident in Norfolk. They were the people who first helped me shape my way of looking at the world. My wife, Iris, and three sons, Richard, Charles and Mark, came into my life in time to help with the fine tuning. To all of them I am deeply grateful.

In the University of Hull I am indebted to Professors Allan Patmore and Roy Ward and Dr Derek Spooner for permission to use facilities in the School of Geography and Earth Resources, Mr Keith Scurr, for drawing Figures 4, 5 and 15, Mr Brian Fisher, Mr Alan Marshall and Mr R. Wheeler-Osman for photographic assistance, and Mrs Pat King for help with providing maps.

In Stibbard, Norfolk, Mr John Spencer Ashworth and Mr Jack Loades for reviving memories of sixty years ago, Mr Thomas Cook for permission to photograph the grounds of Sennowe Hall, Mr and Mrs K. D. Hurrell, for permission to revisit what is now the 'Old' Rectory, and Mr D. R. G. Daniels for help in correspondence.

Further permissions to reproduce illustrative material are acknowledged in the captions to various Figures.

Valuable suggestions were received from the following, who kindly read the manuscript: Professor Gordon Orians, Professor Grant Hildebrand and Mrs Naomi Pascal, all of the University of Washington, and Professor Kenneth Helphand of the University of Oregon, U.S.A., Dr Jane Gear of Cottingham, who also drew Figure 2, and not least Dr Jacquelin Burgess, Reader in Geography at University College, London, who nobly read the manuscript twice!

My thanks also to the University of Hull Press, and to others unnumbered who, though making a contribution in whatever way, must rest content with a gratitude none the less sincere for its not being specifically recorded in these pages.

LIST OF MAPS AND ILLUSTRATIONS

Frontispiece. Jay and Iris Appleton in Cottingham, 1990. Photo by Charles G. Appleton.

PART I THE WORLD MAKES ME

1 AN ENGLISHMAN'S HOME

'Heaven lies about us in our infancy.'

Antecedents

So the train mov'd slowly along the Bridge of Tay,
Until it was about midway,
Then the central girders with a crash gave way,
And down went the train and passengers into the Tay!
The Storm Fiend did loudly bray,
Because ninety lives had been taken away,
On the last Sabbath day of 1879,
Which will be remembered for a very long time.

Thus, in one of his more excruciating poems, does William McGonegall recount the story of the Tay Bridge Disaster. On the night of 28 December 1879, the bridge collapsed and a train fell into the ice-cold waters of the Firth of Tay with heavy loss of life and, thanks to McGonegall, people have been laughing themselves silly about it ever since.

Forty years later, to the day, a curate's wife in Leeds, England, celebrated the anniversary of this tragicomic event by having a baby, that is to say, me. I sometimes wonder whether my love of trains, of Scotland and of the absurd is in some macabre way connected with this coincidence, but I shall certainly resist the temptation to argue this thesis now. Suffice it to say that the suburb of Headingley, a name known to cricket enthusiasts throughout the world, was to be the scene of my first adventures in environmental perception.

I still have vivid memories of the tall Vicarage and its attic tenanted by the impoverished curate, of being bathed by my mother in a white enamel washing-up bowl in front of the open fire and of being pushed down Shire Oak Road in a high pram, intermittently throwing out my rattle and exclaiming 'Motor car!' whenever I saw what, in those days, must have been a comparatively rare sight, though, I suspect, less rare in Headingley than in most other suburbs of Leeds.

I now know that these memories, vivid as they remain, were memories, not of these incidents as they occurred, but as my mother recounted them to me years later. The fact is that, at the age of eighteen months, I left Leeds for East Anglia and never re-visited these early haunts for another twenty years. When I eventually saw Shire Oak Road and St Michael's Vicarage again, neither of them bore the slightest resemblance to those pictures I had cherished so vividly in my memory. The Vicarage had even moved itself to the other side of the road!

I know that many of my habits and tastes must have started to be moulded by those first contacts with the real world in Headingley, but I know it from general principles, not from specific evidence. The little toddler who arrived in North Norfolk in the summer of 1921 to be exposed to the influence of a very different environment must already have shown highly individualistic patterns of behaviour. How these patterns might have developed had we continued to live in Yorkshire there is no means of telling, neither is there any profit in speculating. It was in these new surroundings that I was destined to spend the most influential formative years when, possessed of the power of locomotion and able to wander for the first time beyond the watchful eye of my mother or her authorised representative, at least for short periods, I was able to embark on the serious exploration of the real world and to abstract from my discoveries what I needed for the creation of its counterpart, that private picture of reality as I conceived it myself. This new home and its immediate surroundings were to provide the principal exercise ground where I was to learn how to grow eventually into the person I now am, and for this reason it merits a more detailed description than any of the arenas in which the later scenes were set.

Before embarking on that, however, I must tell you a little about my parents and their respective backgrounds, because, if any people have a place in this story, they must have the strongest claims. I can now see that many of their tastes and preferences, and above all their sense of values, powerfully influenced my own, and I shall refer to this again at the appropriate times. They, presumably, were also children of their own early environments, and in some instances I have been able to see evidence of this, particularly in the case of my father, probably because I came to know the places associated with his early years better than my mother's childhood haunts.

My father was christened James Ashley, but he was known to all his closer friends and relations as Tom. He came from a family whose professional interests seem to have alternated for some generations between the Church and the Law. His father, son of the Vicar of Worksop, Nottinghamshire, had established a firm of solicitors, Wright and Appleton, in Wigan, Lancashire, in the middle years of Queen Victoria's reign, and his two older sons, my father's brothers, joined him in the firm and eventually took it over, leaving my father to seek a livelihood in the Anglican Church. My grandfather died in 1925 when I was five. (One of the consequences of being born when I was is that I can easily calculate how old I was at any given time by taking the last two digits of the year and subtracting twenty. This only fails to work on the last four days of the year.) I'm told that he was kindly disposed towards my sister and me, though I confess I remember him only as a beard in a photograph.

My paternal grandmother I remember much more clearly, since I had reached the age of twelve when she died. Neither of my grandparents is of immediate importance for our purposes. I have no doubt they must have exercised an influence over my father's tastes, as he did over mine, though I should be unable to explain exactly how. But if they are not strictly relevant as people in the present context, the house they lived in is crucial. It was called Bradley Hall and you will be hearing more about it.

My father had no great academic ambitions when he entered Keble College, Oxford, in 1906. He aimed at, and successfully obtained a Pass Degree, having spent most of his time in the pursuit of sport. He arrived there with a reputation, acquired at Shrewsbury School, as an oarsman and a footballer. 'Football' at Shrewsbury meant Soccer, and he immediately obtained his Blue as a goalkeeper in which capacity, however, he broke both his wrists. That was the end of his rowing career, but he continued to keep goal for the University in each of his three years as an undergraduate and went on to become the regular goalkeeper of the Corinthians in the days when, as the leading amateur club in England, they were able to hold their own with any of the professional clubs in the Football League and the F.A. Cup. He relished every sport he took part in, reached county standard in hockey, playing centre-half for Shropshire, and played to a handicap of four on the golf course even with a pair of wrists which never fully recovered from the fractures. A football also broke his nose and this had permanently changed his appearance before I ever saw him.

After Oxford he spent a year teaching in a preparatory school in Shropshire (it was then that he played hockey for the county), before going to Leeds Clergy School. He was ordained into the Anglican Ministry in Durham Cathedral and took up his first curacy in Sunderland close by, where he became acquainted with the family of one James Ayers, Chartered Accountant, and later married the eldest of his three daughters, Lilian. My mother was seven years older than he, though I didn't know this until I was at least in my teens. Her training and experience as a schoolteacher certainly had an effect on the development of my awareness of the things around me, and I never grew out of the deep affection which a small child normally feels for its mother.

I never saw my maternal grandfather, who died before I was born, but my grandmother survived until I was into my twenties and in fact lived with us for many years. To distinguish her from my other grandmother we called her Granty. She was short, rotund and good fun. One of our favourite pastimes was to reduce her to helplessness by provoking a fit of laughing. This was extremely easy and invariably successful. I can see her now sitting there and quietly shaking like a jelly, tears streaming down her face, and unavailingly trying to utter words of reproof to tell us what rotten little monsters we were. We were cruel, but how she loved it!

I believe my parents' marriage was a deeply happy one. If they ever quarrelled we never saw any sight of it, and I have no doubt that their attitude to my sister and me and to each other had much to do with the establishment of my concept of 'home' as a very firm pivotal point in my mental map of the world around me. Their marriage had been delayed for some years by the First World War. My father had become an army chaplain, serving in Macedonia, Imbros (now called Gökçeada in deference to Turkish political sensitivity), an island in the entrance to the Dardanelles, and Egypt, before being invalided out with malaria. The former Principal of the Leeds Clergy School had, in the meantime, become the Vicar of Headingley, and this was the connection which led to my father becoming his curate and living in the attic of his house. But the malaria didn't end with hostilities, and he went to Headingley as a sick man. It was on medical grounds that he was advised to leave the atmosphere of the town and take up a job in the country, and that was how we came to find ourselves in the Rectory at Stibbard, a parish of 373 souls, according to the 1921 Census taken a couple of months before we arrived, in a fairly isolated part of North Norfolk.

In his new surroundings my father's health generally improved, and when he was well he was a picture of fitness. He played for Stibbard in the local football league to the delight of his parishioners but the dismay of his opponents from neighbouring villages who had not been accustomed to having to put the ball past a goalkeeper who had narrowly missed an England cap. But every few weeks this condition of health and vigour was rudely interrupted with monotonous regularity by 48-hour bouts of the fever. The attacks were predictable, and we had to build our plans round this pattern of life which lasted for approximately fifteen years. Then, one day, when an attack was expected, it failed to arrive, and my father lived for another twenty years without a single recurrence. I believe, however, that this experience tended to reinforce the symbolic significance of 'home'. Just as a wounded or sick animal makes for its preferred place of refuge, so my father made for his bed; not just any bed but his own, where he could be tended by my mother and left alone to look forward to the day after tomorrow. In retrospect I think something of this feeling must have brushed off on to me. It helped to build up the image of Stibbard Rectory as the home base, or, if you like, the harbour from which one ventured forth to sample the world beyond, but always in the hope and expectation of a welcome return.

The parish of Stibbard lies some four miles east of the little market town of Fakenham in North Norfolk, half a mile off the Fakenham-Norwich road and about ten miles inland from the coast near Blakeney. It's located in an undulating landscape of relatively low relief, by which I mean that, while the higher parts of the parish rise a little above the 200-foot contour (about 60 metres), the lowest extremity is not much below 150 feet. This is, however, very different terrain from the Fenlands, which is what many people erroneously think of as the typical East Anglian landscape. I shall have more to say about the details of this little piece of the English countryside, but it will be best to start with the Victorian residence which stood in the middle of it and which was to be home for me until the age of twenty.

There's a well-known saying that an Englishman's home is his castle, and this imagery will do very well as the basis for a description of Stibbard Rectory and its grounds. It may seem far-fetched to employ the language of military architecture, but in the child's world of make-believe the idea of safety and security can find expression through any symbolism that seems to work. The image of the castle as a romantic development of the place of safety is fostered in fairy tales and not least in children's illustrated books. The ambivalence of such a place is made clear in such stories. The castle itself may be inhabited by a prince or an ogre. It may be a place where visitors, including ourselves, will be made welcome; it may be a place from which, as intruders, we should expect to be repulsed. A fundamental objective of environmental exploration is to provide ourselves with the knowledge which will enable us to make an appraisal of the places we explore in precisely these terms. Are they nice, comfortable, homely places, or are they hostile, forbidding,

unsympathetic and alien? Such a primary classification is basic to
the behaviour, not only of young children, but of the young of other
species as well. In a more hostile world than that in which I was
brought up the distinction might be quite literally a matter of life and
death.

The Keep

Let's begin, then, with the castle keep, that inner sanctum in which I
spent the greater part of my time. The house itself was not large in
comparison with the houses of some of the neighbouring clergy, a
blessing for which my mother was truly thankful. That it was much
larger than most of the other houses in the village was a fact which I
took for granted. There were one or two farmhouses as large as ours
but for the most part my father's parishioners were a congregation of
cottage-dwellers. They belonged, however, to the world outside the
castle walls of which more anon.

An estate agent's description of Stibbard Rectory (Figure 1)
might have said:

> Detached country residence with entrance hall; three
> reception; large kitchen with solid fuel range; pantry;
> larder; back kitchen; toilet; large utility room; cellar;
> and on the first floor four principal bedrooms;
> dressing room; two back bedrooms; toilet; and on the
> second floor one bedroom and a bathroom. Two
> ranges of outbuildings comprising coal shed, wash-
> house, outside toilet, workshop, woodshed, small
> stable, garage and general purpose shed with loft
> over. Large garden, tennis court and orchard.

I use an up-dated terminology. We had never heard of a utility room,
and what sort of a range was there other than 'solid fuel?'

I will not conduct you through the house room by room, and I
give this catalogue only to show the sort of place I came to regard as
comprising my 'personal space'. The usage of these rooms under my
mother's *régime* was not quite as described above. One of the
reception rooms was my father's study; one of the principal
bedrooms was our playroom, known as the nursery until my
grandmother (Granty) came to live with us, after which it became her
bedroom. Of the two back bedrooms one was known as the lumber
room until my father bought us a small billiard table which would
just about fit into it with enough room to walk round it if not to play
a shot. We had a little sawn-off billiard cue for stabbing our way out
of trouble if we found ourselves cushioned at either end of the table.

FIGURE 1. STIBBARD RECTORY, NORFOLK, FROM THE WEST. This is a pen-and-ink sketch which was made by the author in March, 1940, shortly before leaving Stibbard Rectory. The window to the left of the front door is the dining-room and the spare bedroom is above. Helen's bedroom window is above the front door. The french window (right) opens out of the drawing-room and my parents' bedroom window is above. My bedroom is to the right of that and my father's study is below it. The coal shed appears to the left of the house. The apple tree on the lawn has been omitted. See also Figures 2 and 4.

FIGURE 2. BIRD'S-EYE VIEW OF STIBBARD RECTORY. View from the west showing part of the garden, churchyard, etc. The drawing, by Jane Gear, may be read in conjunction with the sketch-map (Figure 4).

It's a miracle we never cut the cloth. But that belongs to a later period. In this first exploratory phase it was the lumber room and very exciting!

The smallest of the principle bedrooms was allotted to me, while my sister had to make do with my parents' dressing-room. She experienced a perpetual problem something like the one I have described in the billiard room. It was just as well that she was more tidy and orderly than I, as she had nowhere to put anything, but she didn't seem to mind. She was thirteen months younger than I and my constant companion in the great adventure of learning about the world. She was slanderously alleged to have a fiery temper on the grounds, I assume, that red-headed means hot-headed. Even when I called her hair red, which I wasn't supposed to, ('auburn', dear!), or, worse still, warmed my hands on it, she never displayed more than a justifiable irritation. I don't remember ever having a serious tiff with her or ever refusing to share a secret, except about Christmas presents and other objects of legitimate deceit. Illegitimate (i.e. real) deceit I occasionally practised against others, but more often than not she was party to it. The partnership was not simply one of convenience; I was deeply fond of her and could never imagine how a brother and sister could fall out for more than a few moments. She was (and is) called Helen Mary after my father's eldest sister who had died at the age of nineteen.

You will have gathered from my brief description of the house that it was big enough and rambling enough to provide a promising arrangement of interior space for the imagination to work on. 'Downstairs' was more or less a public building. My father's study was his parish office and one never knew whom one might meet in the hall on his or her way there. Not only that, but most of the village organisations, such as there were, were liable to meet in the drawing room. It was the only place there was for the Mother's Union, the Parochial Church Council, the Committee of Stibbard Football Club and many others to meet, and nothing seemed more natural than that our family should share its living-space with the parish. This, I dare say, was a contributory reason for the importance which I attached to the privacy of my own bedroom, which was large enough for me to have my own desk in it from quite an early age, and as I moved into my teens I valued it increasingly.

But it was the nursery which I now think of as providing in those early days the solution to the problem of 'territoriality.' The nursery was a fairly spacious room and it had three built-in cupboards at one end. Since they occupied the space under the attic stairs they diminished progressively in size. That on the left was big enough for any adult to enter comfortably. The middle one was rather more restricted, while that on the right was so tiny that only

Helen and I could actually get inside it. Between them they offered enormous scope for every kind of inventive fantasy. They were pressed into service as rabbit-holes, prisons, captain's quarters when the rest of the room was the deck of a pirate ship, green rooms for the production of plays, *real* hiding places providing a refuge from unwelcome visitors, dark places for the weaving of spells and the performance of magical rites and all kinds of sorcery; in short, they were invaluable items of educational equipment for the cultivation of the imagination.

Another room which had a charisma of its own was the bathroom. It never struck me as at all peculiar that one should have to go up to the attic to take a bath. If anything was strange it was the notion that the bathroom should be on the same level as the bedrooms, as it seemed to be in most other houses I visited. *They* were all out of step. The bathroom was a narrow, elongated room with a window at one end and another deep, dark cupboard at the other. It had a large galvanised iron tank perched just below the ceiling at the foot of the bath. I well recollect how this tank had to be filled by a long-handled pump in the back kitchen until my father installed a paraffin engine in what I have euphemistically called the utility room. It did have a name but curiously enough I can't recollect it. I must remember to ask my sister.

When I say my father installed the engine, I don't mean he engaged an engineer or a plumber to do the job. He was fascinated by any practical challenge and never deterred by the fact that he had had no training or experience in the particular skill required. By a combination of enterprise, effort and natural ability to understand how things work he successfully accomplished all sorts of improvements to the home. Although the expression had not then been invented he was the D-I-Y man *par excellence*. He fitted a complete coke-fired central-heating system in Stibbard Church with large-bore pipes, digging a subterranean boiler-house and doing all the bricklaying as well as the plumbing almost single-handed. Home improvements, therefore, were by comparison no problem. The most ambitious enterprise he undertook in the house was the installation of an acetylene gas lighting system which superseded the oil lamps we had during the first years of my life and which I can still remember suspended by chains from the ceiling. The acetylene plant was just like a miniature gasworks with a cylindrical gasholder floating on water and a pretty authentic pong at close quarters. It was housed not far from the back door in a little brick shed which always had a heap of white sludge beside it - hydrated calcium carbide, the waste product of the acetylene process.

Perhaps the only other room which calls for special mention was the cellar. Frankly it was a disappointment, almost a fraud. It was so

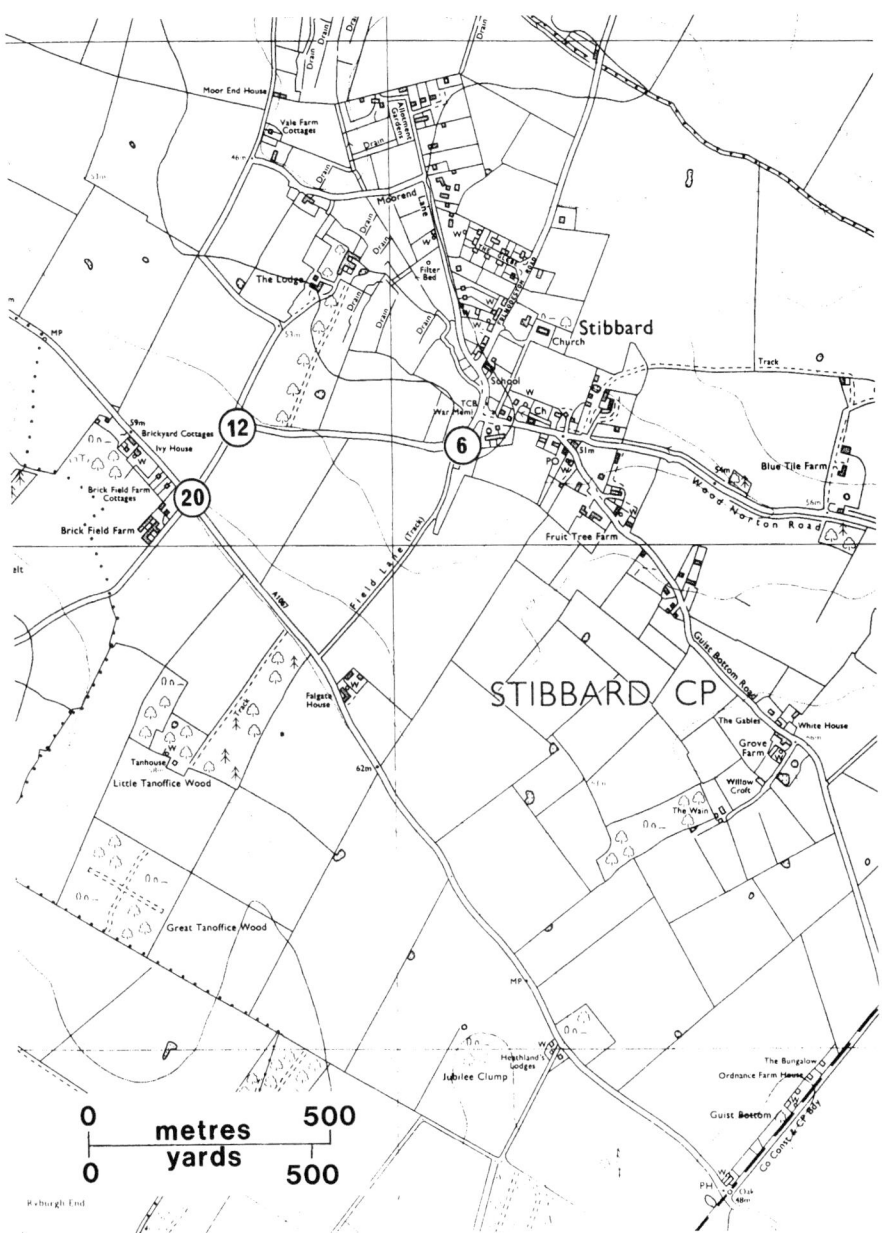

FIGURE 3. DETAIL FROM ORDNANCE SURVEY MAP, 1:10,000 SHEET TF92NE.
This detail is taken from the 1985 edition and contains a number of houses built since
the nineteen thirties. Stibbard Rectory is situated just to the west-north-west of the
church. The A1067 road is 'the turnpike'. The numbers 6, 12 and 20 have been
inserted at the road junctions corresponding to the angles in the number-sequences
shown in Figure 5 as subsequently explained in the text (pp. 22-3). Reproduced from
the 1985 Ordnance Survey 1:10,000 map with the permission of the Controller of Her
Majesty's Stationery Office, Crown copyright.

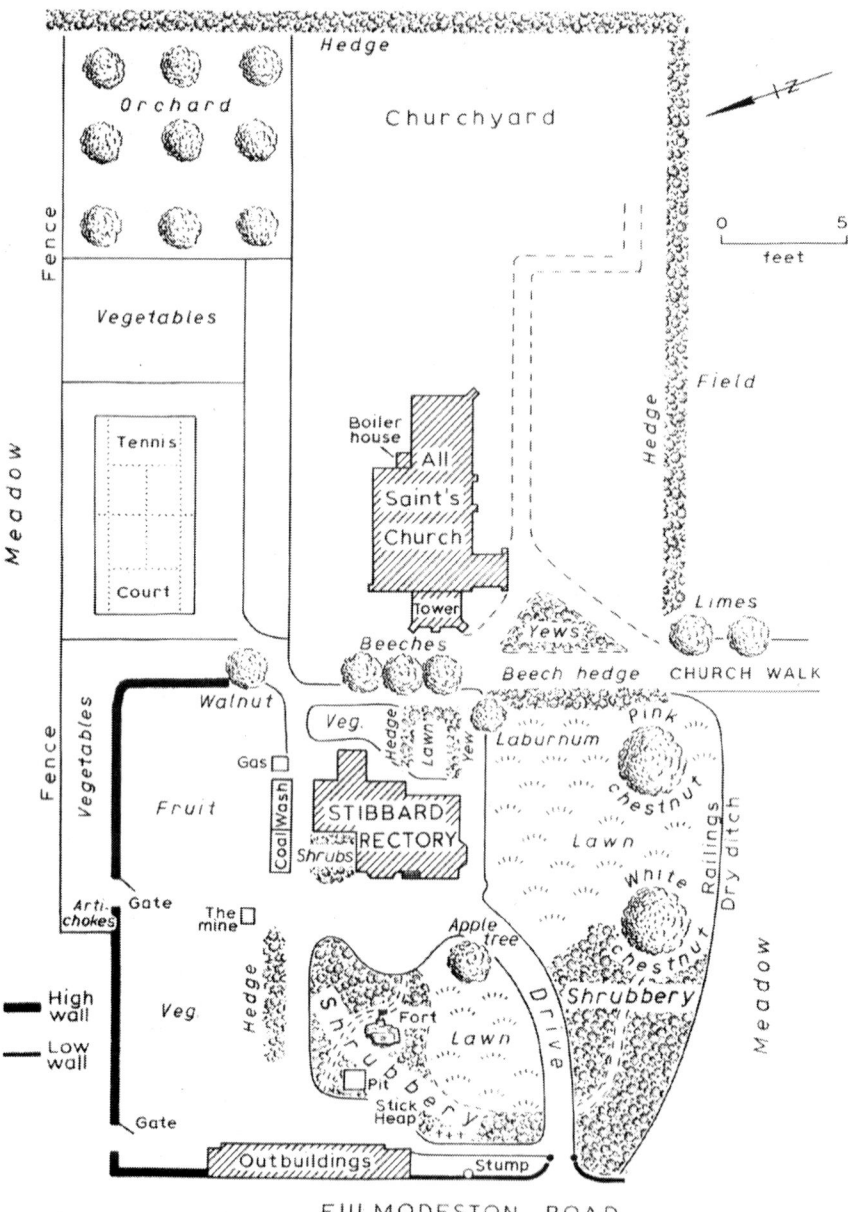

FIGURE 4. SKETCH-MAP OF THE RECTORY GARDEN AND CHURCHYARD, STIBBARD. Scale approximate. Drawing by Keith Scurr.

shallow that three or four steps sufficed to reach the floor. It didn't go *under* anywhere except the billiard/lumber room on the first floor so it had a very high ceiling. No cellar like that could be anything but a failure, but, shallow as it was, as if to compensate for its inadequacy, it got flooded every winter, and that was fun. I suppose it was also a pretty good reason for not making it any deeper, since it was virtually useless for anything except storing eggs preserved in a huge earthenware pot of waterglass. How I envied neighbours with proper cellars which actually went right under their houses!

This, then was our innermost refuge, our enclosed living space, the keep of the castle. From its windows we looked out over the surrounding terrain as a medieval potentate might have looked out over his principality. And when we looked out, what did we see?

From my bedroom window the view was to the south, or nearly so (Figures 1-4). I shall refer later to a problem caused by the fact that the house was not set quite in conformity with the points of the compass but we'll leave that for now. The picture I saw was framed by two splendid horse chestnuts which in turn were flanked by screens of other trees, mostly deciduous and therefore opaque for only half the year. Between the chestnuts there opened out a prospect over the centre of the village in a shallow valley a quarter of a mile away. This was marked by the village green, a farm, a shop and the village pub. The other hundred or so habitations were scattered in isolation or small groups over a much wider area. Half a mile beyond the green there stretched out a skyline along which lay the Fakenham-Norwich road, made by one of the old turnpike trusts, probably in the eighteenth century, and still called 'the turnpike.' It was a bare skyline apart from a few isolated trees, two of which, one larger than the other, merged in *silhouette* and assumed the shape of a round rabbit-like creature which, year after year, munched away at the hedgerow it surmounted. I used to say goodnight to it last thing before I got into bed. I believed it to be generally benevolent, though there were times when it assumed a more sinister character - just like my other friends, I suppose. Curiously enough I don't believe I ever gave it a name.

The skyline was slightly higher than our house and therefore looked like the end of the world. It changed its mood with the changes in the sky, and I'm sure that, at a very early age, I was aware, first, that the sky is an integral part of the landscape, and, secondly, that, in looking at a landscape, we are always conscious that it comprises more than we can actually see. Although I knew that beyond it the ground sloped gently down into the valley of the River Wensum, and although I could at least partially supply from my growing knowledge of the adjacent parishes the details which I couldn't actually see, these details had always to be fitted into a pictorial composition by the imagination. There was therefore a constant challenge, never to be resolved, between the seen and the unseen. Only much later did I rationally comprehend the importance of horizons in the aesthetics of landscape, but I had certainly experienced their magic in early childhood. That horizon was the first I ever knew intimately, so to speak, the prototype of a whole class of emotionally important visual experiences.

I always used to think that, just as Helen had to put up with more cramped conditions in her bedroom, so she was less fortunate in the view from her window. Any possibility of a distant view was frustrated by the trees in the garden. The one compensation was the sunset when the glow of the western sky was strong enough to force its way through the foliage; that set up even more powerful speculations about what lay beyond the horizon. In summer the sun set behind the ashes and sycamores which fringed the lawn, but in winter it sank to rest behind a little clump of trees in the far corner of the shrubbery beside the front gate among which a shapely pine was the most conspicuous. Its effect on me was profound and I haven't the slightest doubt that, reinforced as it was by innumerable subsequent experiences, it marked the beginning of a taste for conifers in general and pine trees in particular.

Apart from the kitchen, the back kitchen and the nursery there were no rooms which faced east, though there was also one window on the stairs and another in a passage. The view from all of them was blocked by a row of splendid beech trees with smooth trunks and graceful foliage, and behind that the west end of the church. By East Anglian standards it was quite a small church, but the tower stood so close to the house that it seemed to overshadow it, giving a quite dramatic meaning to the expression 'towering above', particularly to the eyes of a very small child. The effect of this opaque screen was to turn the eyes downwards from the windows to the diminutive lawn horseshoed by tall evergreen hedges. I was over fifty before I attempted to rationalise this in a theory of landscape aesthetics, but as early as I can remember I was keenly aware that the view from this side of the house, enclosed, protected, almost even secret, was poles away from that from my own bedroom which invited the eye to pass over the more sheltered foreground and enjoy the sense of space beyond.

The most spacious view of all, however, was to the north. There were only two windows in the house which faced in that direction and they enjoyed the additional advantage of being on the top floor. In a middle-class household of even modest means a resident maid was by no means unusual in the nineteen twenties and thirties, and one of these windows was therefore strictly out of bounds. The other was in the bathroom, and in some ways that was our most exciting window on the world. There is very little of Norfolk which rises above the 300-foot contour, but one such patch lies between Thursford and Melton Constable, and this formed the northern skyline some four miles away. In contrast to the familiar horizon which I looked out on from my own bedroom it was well-wooded and gave the appearance of being the edge of some vast forest which went on for ever and ever. Even when I had come to know that this was not so, I cherished the illusion.

Much closer than this, indeed little more than a mile away, was an outlier of these woodlands known as The Severals. It was a mixed stand of deciduous and coniferous species and it contained a few exceptionally tall conifers which projected like spires above the general level of the canopy. These, too, were the forerunners of what was to become an important type in my system of preferences.

The middle distance of this landscape was filled by gently undulating ground quilted in a pattern of hedged fields typical of the mixed farming of the district. Here and there was a barn, a cottage, even a farmstead, but there was not much habitation between our house and Fulmodeston, a couple of miles away. The most compelling part of the view, however, could only be attained by climbing on to the washbasin, putting one's head out of the window and craning round to the left, that is to say to the north-west. If one could find the physical stamina to stay there long enough in this precarious posture one might eventually be rewarded with the sight of a train on the Midland and Great Northern Joint Railway, also a couple of miles away. Remote and inconspicuous as it was, its symbolic significance would be difficult to exaggerate. In the first place it reassured us that this little corner of England at the back of beyond, off the beaten track and almost off the map, was still a part of civilisation, sharing with the rest of the world the engineering wonders of the nineteenth century if not yet, not quite yet, the twentieth, thereby achieving the best of both worlds - isolation without separation, seclusion without loneliness. And in the second place it was a kind of signpost to everywhere else, a messenger, like Noah's dove, arousing speculation about what those other places were really like, far, far away.

The Inner Bailey

Every good castle has a good view. It's a fundamental part of its system of defences. But it has other defences too; typically an Inner and Outer Bailey. These are the immediate adjuncts of the fortress, the outdoor components of its protected space. Whatever the distant prospect, from every window in our castle the Inner Bailey formed the foreground. Here we could savour the excitement of a close contact with nature. We could explore her, test her, find out how far we could go with her, and all within the security of the ramparts. It was a wonderful stage on which to play the part of a growing child.

If you look at my little sketch map (Figure 4) you'll see that on more than two sides the garden was enclosed by a brick wall. Along the northern side it must have been some seven or eight feet high, though it seemed much higher. This high wall was continued for a short distance along the western side and was then succeeded by a group of outbuildings to which it was connected by a kind of seamless join so that, from the road, there appeared to be a single windowless wall running from the corner of the garden to the far end of the buildings. Beyond this a much lower wall extended as far as the entrance to the drive. The entrance in turn was flanked on the south by a thick plantation of trees with shrubs under the canopy. It contained laurels and snowberries dense enough to provide effective concealment, while the higher trees, including the pine, which I mentioned before, and an oak, but mostly sycamores, roofed the whole thing over. From the little pathway which led through the middle one could command a view of the adjacent meadow to the south without any chance of being seen.

Next came the two horse chestnut trees, first the white one, then the pink one, and the vitally important gap between them where the lawn came up to the meadow. This gap was spanned by a length of wrought-iron fencing, its bent bars testifying to the hours my sister and I spent standing on them as we gazed down the gently sloping meadow to the little stream at the bottom and the village green beyond. When we were playing on the lawn the width of this gap between the chestnuts was enough to keep us constantly aware of the openness of the meadow. There were many places where we could peep out of the garden from positions of concealment, but this was the only place where others could effectively peep in. Not that they ever did; but that was not the point. It was as though two different kinds of space came together and mingled in this gap, entering and exiting like the ebbing and flowing tide. The price we had to pay for the view was the slightly disturbing sense of exposure, not to any real danger, of course, but rather to some imaginary invader of our cherished privacy. Beech hedges and beech trees and a short resumption of the high wall on the eastern side completed the circuit and closed the defences.

The garden was full of fascinating places and interesting things. My mother's favourite part was the front lawn (between the house and the two chestnuts, Figure 4) which contained the principal flowerbeds. I was never particularly interested in these and I should have been hard put to it to name any except the most common flowers which grew in them. In fact my most persistent memory was of a clutch of tiny chicks in a wire-netting pen in the middle of the lawn. The old hen was incarcerated in a little coop from which the chicks could come and go as they pleased by squeezing between the vertical wooden bars. Unfortunately the coop was just the right height for jumping off, and I can still see a hapless little ball of yellow fluff staggering off, fatally injured by my having inadvertently jumped on it, to expire at the feet of its mum. No doubt I stopped crying after a few minutes, but I never passed the spot again without a twinge of melancholy and a sharp pang of guilt. It was as though the spirit of the dead creature had taken up perpetual residence at the scene of the crime to torment its assassin every time he had the temerity to pass that way. I think this was perhaps the first time I ever realised how powerful can be the association between places and the events which happen in them. Just as one's visual memory of events incorporates the scene in which they occurred, so one's image of places can be distorted by episodes which logically should become irrelevant as soon as they are over. But logic is a stranger to many visual experiences even of an adult, never mind a child.

On the left-hand side of the drive as one entered it from the front gate a small lawn sloped gently down from the house towards the road. A Keswick Codling apple tree stood on it in splendid isolation like a lighthouse on a promontory. It bore juicy sharp apples, yellow and quite soft, with a kind of suture down one side as though the skin had been sewn up to keep the juice in, and it was a good tree for climbing. I liked this lawn the better of the two. It was irregular in shape and fringed by a curving curtain of foliage belonging to the various bushes behind, part of another little shrubbery-wood with, again, a little path passing through it in a sigmoid curve. It was a place full of all sorts of excitement. In the very middle of it we constructed a fort, a kind of trench covered with old linoleum and other makeshift materials whose more orthodox domestic careers had come to an end. It commanded a view over the Pit, which I'll tell you about in a moment, to the little range of outbuildings at the bottom of the garden. On that side its roof was raised a few inches above the level of the surrounding ground by being placed on a low earth bank containing a couple of apertures. One was for looking through, the other for firing the gun. A toy gun, of course. Oh no! By no means!

The gun consisted of a short length of four-inch cast iron pipe, about three feet long and open at both ends. About half of it was inside the fort, the other half projecting in the direction of the Pit. A few morsels of calcium carbide, filched from the gas plant, were put into a Lyle's Golden Syrup tin - the green one with a picture of the lion and the bees and the quotation from the Book of Judges - together with a tablespoonful of water. The lid was replaced firmly and the tin was inserted, lid first, into the pipe from the end inside the fort. Before this, however, a small hole had been punched in the base of the tin through which a fuse had been introduced consisting of a piece of string soaked in paraffin or candle grease. The fuse was lit and the occupants of the fort retired to a safe distance, waiting for the lid of the syrup tin to fly over the Pit and bombard the woodshed door. The occupants consisted of my sister, myself and my friend John Ashworth, of whom more anon.

In retrospect the whole exercise seemed so improbable that, when I first drafted these lines, I decided to omit any reference to this gun on the grounds that my memory *must* have played me false, or, if not, I should in any case have scant hope of being believed. But before finalising my text I paid a visit to Stibbard where I encountered John Ashworth for the first time in forty-five years, and he, unprompted, recounted a description of the contraption exactly as I have just given it. I have therefore decided to restore it to the text, not least because, as you will discover, the military theme plays a persistent if highly ambivalent role throughout this book, and with hindsight I can see much significance in this early experiment. Playing soldiers is, of course, one of the most popular of childhood pastimes, but I'm astonished how we carried it to such a level of sophistication and technological enterprise. How my mother, who would have been horrified at such an operation, could have failed to be aware of what was going on I can't imagine. It made a dreadful noise! Presumably we always made sure she was out. I've already told you that Helen and I were not beyond indulging in the occasional act of conspiratorial deceit.

Not least it must have occurred to me that, with all that volatile gas hanging about in such a confined space, the thing might well have gone off prematurely while we were still inside the fort. The motivation must have been very powerful to overcome so many sound reasons for not doing such a stupid thing. I sometimes wonder whether the device was unique in the history of artillery. Was there ever another acetylene cannon?

But to revert to the Pit. This was a square brick-lined hole in the ground into which was deposited all the household's biodegradable waste from potato peelings to lawn mowings. In summer the upper layers tended to dry out. Then they became firm enough for us to jump up and down on them, a sort of trampoline, if you like, but if the crust was too thin there was a ghastly squadgy mess underneath, and I well recollect being dunked in the bath after I had misjudged it.

Next to the pit was the stick heap. It had rats nesting in it, a fact which I believe we successfully concealed from my mother in case she should order their professional extermination. It stood immediately opposite the little shed at the end of the range of outbuildings and the loft above was probably the first 'hide' from which I had the opportunity of observing wildlife. I can't say I ever acquired an affection for the rat, but I remember thinking how safe and cosy it must be to have a nest somewhere in the middle of that pile of sticks and somehow I developed a kind of empathy with this not-very-attractive creature, not so much for itself as for its nesting-place.

Beyond the stick heap a little finger of shrubbery divided the lawn from the path which led to the front gate. It was just about wide enough to provide decent privacy for the tiny tombstones which marked the last resting-places of various rabbits and guinea-pigs which we had successively cherished and for which we had expressed our affection by keeping them imprisoned for life. I often wondered whether they or the rats in the stick heap lived the happier lives.

Between the little path and the road stood one of the most exciting objects in the Inner Bailey. It was the trunk of a more-or-less dead tree which at one time must have been pollarded to produce a multiplicity of branches. Most of these had dropped off leaving a kind of cauldron-shaped hollow at the point where they had diverged some seven or eight feet above the ground. In this cauldron we could lie totally concealed from the public traffic on the road, which, however, we could observe the more easily because the trunk itself leaned out over the low garden wall to command a superb view both ways along the highway. On at least one occasion we took up a strategically advantageous position very early in the morning equipped with suitable ammunition. Opinion differs as to whether this was a bucket of water or a bag of flour. The target was an open lorry taking men to work and the theory was that, by the time the driver in his cab had been appraised of what was happening, the lorry would have gone too far to permit a retributive strike.

There were innumerable other features of the Inner Bailey which provided scope for imaginative exercises. I shall mention only two. The first was the yew hedge flanking the gravel path which led to the churchyard. This was a very secret place. For some years it contained a tree house which we erected in it and which was virtually invisible from ground level. I spent many hours there. The second was a hole in the ground known alternatively as 'the Dug-out' and 'the Mine'. It was much deeper than the Fort, nearly six feet, I should say, and more or less square. It also had a roof made of whatever material had come to hand - an old door, a piece of corrugated iron and much sacking. It had a fireplace and an oven made out of a large square biscuit tin. A length of cast iron pipe of the same specification as the gun served as a chimney; in fact, come to think of it, it probably was the gun re-cycled to serve in a civilian capacity after being banned from further military use. The Mine, however, was a rather later teenage addition to the inventory of garden follies.

The outbuildings provided a further range of opportunities for adventure. They were made of old, mellow red bricks and they were full of dust and cobwebs. The little stable contained an iron manger, a sort of semi-circular hay basket screwed into the wall with a name-board affixed proclaiming that it belonged to Natty Bell. Natty Bell was the obstinate little donkey which we took over from the previous occupant when we moved in, and which, with the trap, provided our only means of locomotion. He must have been called after the prize-fighter in Jeffery Farnol's recently published novel *The Amateur Gentleman*, which no doubt said something about the temperament of the little donkey. I can just remember being driven to a party at a farmhouse in Fulmodeston behind Natty Bell, but it's a hazy recollection, and it may be no more authentic than those 'vivid' memories of Headingley.

You may well imagine that the Inner Bailey contained everything a young child could ever want. But the urge to explore fresh fields is a powerful one especially when coupled with the desire for the acquisition of territorial rights, and the Outer Bailey therefore also had an important role to play in this game of environmental interaction, so I shall tell you a little about that.

The Outer Bailey

The Outer Bailey consisted of five distinctive parts. They acquired a certain unity from the fact that they all lay outside the wall or the hedges which I now believe, with the benefit of later experience, must have delimited the boundaries of the original garden and to which the garden of the Old Rectory now seems to have reverted. Four of them were legally just as much a part of our garden as the Inner Bailey. They appear to have been the fruits of later annexations made out of a single adjacent paddock, latecomers to the club which somehow never seemed quite to achieve full membership status. They had a sort of extra-mural feel about them, which was appropriate enough, I suppose, for those parts which lay literally outside the wall, and they were protected from the north only by a line of posts and a few strands of barbed wire (marked 'Fence' in Figure 4), paltry protection indeed from the winds which blew down from the Arctic. Don't forget we were at nearly fifty-three degrees north latitude.

The first of these areas was an extension to the vegetable garden. It was, so to speak, wrapped round the north-east corner of the garden wall. It grew artichokes, and you can *hide* in artichokes if they are of the right kind. On the whole, however, it was a featureless piece of ground and the least interesting part of the whole domain. It was the customary place of execution for chickens destined for the table and I found it altogether an uncomfortable spot.

The second of these extra-mural units was the tennis court. Physically it was equally exposed, equally unprotected, but socially it felt more fully integrated with the Inner Bailey. My father was a very keen tennis player and went to some trouble to maintain the condition of the lawn. I can see him now on his hands and knees, a jam jar in one hand and a kitchen spoon in the other, putting little blobs of sulphate of ammonia on the plantains. The idea that a substance could be lethal in concentrated form but, when subsequently diluted by the rain, could act as a fertiliser for the grass was exactly the sort of thing that would appeal to his economic theories of efficiency through thrift. He and my mother occasionally held tennis parties to which friends from neighbouring villages were invited, and these were memorable exceptions to a generally parochial social routine. Helen and I were encouraged to serve lemonade from glass jugs covered by fine-mesh nets with little glass beads dangling round the edges to hold them in place and stop the flies from drowning themselves. Later we learnt to play ourselves, and our growing competence, and consequent interest, invested the place with a new meaning. I never regarded the tennis court as being outside my own territorial space, but its very openness marked it out as different, not strictly comparable with the generally enclosed areas in the more immediate vicinity of the house.

The third division of the Outer Bailey was another patch of vegetable garden, another aesthetically negative area. In fact, until I came to write these lines I had almost forgotten its existence. I think of it now as a space between the tennis court and the fourth division which was basically an orchard, a small plot of grass containing a number of fruit trees and at various times other contributors to our partially self-sufficient economy, namely chicken-runs and beehives. I remember one of my uncles being stung there by a bee, bang in the middle of his balding head, on which the bee, having discharged its venom, became impaled, as it were, by its own sting. We thought this extremely funny and were scolded for not attempting to conceal our amusement, but my unfortunate uncle thought it a perfectly natural reaction and defended us stoutly. There was a lesson in that too, though I'm not sure exactly what it was. Something about Stoicism, perhaps, or the detached objectivity one expects from a lawyer. Anyhow it was a big plus for him. The orchard was bounded on the east by a firmly defined visual boundary in the form of a stout hawthorn hedge with some elegant hedgerow timber. That

really was the end of the Outer Bailey, and no mistake. Beyond lay
the fields and meadows in which we occasionally wandered but not
with the same feeling of legitimacy.

The fifth and final part of the Outer Bailey was different again in
that, unlike the parts I have so far described, it was not a part of the
Rectory garden at all. I refer to the churchyard in which Helen and I
exercised *de facto* rights of access, though we had no more *de jure*
rights than any other children of the parish. We understood from an
early age that certain obligations were expected of us, like not
walking over people's graves, and, subject to the rules, we were
allowed to treat the churchyard as a kind of 'assart', a piece of land
taken in and effectively appropriated for our own personal use, albeit
without title. Nobody ever tried to turn us out and there were plenty
of things to entice us in from time to time.

There were, for instance, the usual churchyard yews, nothing
like as interesting as the yew hedge in the garden, but conspicuous as
churchyard furniture and more than likely to contain a bird's nest.
They flanked the formal walk to the church porch. Then, at the north
side of the church, there was the subterranean boiler-house which, as
I've told you, my father had constructed, and into which one
descended by a vertical iron ladder affixed to the brickwork. The
whole place was charged with the smell of sulphur and the coke
fumes left a slightly acid, metallic taste in the mouth.

But the things I most closely associate with the churchyard were
the slow-worms. I never saw a proper snake at Stibbard, though
grass snakes were alleged to be about. The slow-worm, a legless
lizard, so I was told, was a kind of snake substitute, able to assert
something of the serpent's compulsive fascination without exuding
that sense of evil which has invariably characterised the real thing
since the Book of Genesis cast it in the role of the villain. Just as that
wretched chicken took up a ghostly residence on the front lawn, so
the slow-worms haunted the churchyard, but in the flesh.

These, then, were the castle precincts. They comprised my
personal territory, my patch; and in the evolution of my tastes and
preferences this little fragment of North Norfolk played quite the
most influential part, furnishing the raw material out of which I
made the world, my world, my very own world. The limits of the
Inner and Outer Bailey were what I conceived as the limits of my
defensible space. This space was frequently invaded by my father's
parishioners. My parents conceded a general right of way through
the garden for anyone wishing to go to the church at any time,
thereby avoiding what for some of them would have been an absurd
detour if they were to approach it by way of the long grass-covered
Church Walk. The layout has now been changed by the provision of a
public access road through the meadow in front of the house, but

then it was different. So we had many intruders into our privacy, all, however, exercising a legitimate right of transit and invariably respecting it as a concession.

Beyond the Ramparts

In spite of the passage of migrants to and from the churchyard I had a very keen awareness of the concept of 'inside' and 'outside'. The symbolic interface between them corresponded exactly with, and was reinforced by, the peripheral walls, hedges and fences, and from the inner protected area frequent forays were made into the outside world. But as soon as one emerged on to the road there was a change of mood. The sense of protection was suddenly withdrawn and one was in a public place. *Per contra* the moment one returned into the garden one was emotionally aware of being enveloped by a kind of protective spirit. It was indeed like the feeling of returning to harbour from the open sea.

These excursions into the wider world were made at first, naturally enough, in the presence of grown-ups, but from quite an early age my sister and I were allowed to go alone to fetch the milk, a daily duty which was a pleasure as well as a service. It involved taking a large aluminium can with a lid like a saucepan's down the Church Walk and across the road to the Pound Farm where Mrs Cook or one of her daughters would measure out the correct quantity with a ladle, straight out of the milk churn. This repetitive exercise forged a series of associations, a *Gestalt*, a totality, from which it is impossible to abstract the individual components without altering the way one thinks of them. This is a very common phenomenon in everybody's experience, but I think it was for me the earliest example I can clearly recollect. The aluminium can, the milk, the white gate of the farmhouse, the lime avenue which flanked the Church Walk, the little brick culvert by which it crossed the stream, the rather heavy wooden gate which had to be opened to gain access to the road - all these fitted together like the ingredients of a Constable landscape. To remove any from the picture would be like erasing Willie Lott's cottage from *The Haywain*.

Associations can be of many different kinds and I should like at this point to refer to one which even I find pretty peculiar, like a lot of other things about myself. I believe most people envisage numbers in some sort of pattern. Commonly, for instance, the numbers 1 to 12 may be thought of as they appear on the face of a clock, as in Figure 5, and in this specific context this is how I have always envisaged them myself, except that after midnight the sequence is reversed and proceeds in a counter-clockwise direction until six o'clock in the morning. I suppose that, when I was learning how to tell the time, I never actually saw the clock in the small hours of the morning, so the erroneous impression was never corrected. Having proceeded from my bedtime to midnight up the left-hand side of the clock-face, what more natural than that the hands should retrace their steps in time for me to greet them in the morning more or less where I had last seen them? Regular familiarity with the sight of the clock after midnight in later years has certainly weakened the effect of this pattern, but I still occasionally find myself surprised at, say, two o'clock in the morning to see that the hour hand isn't pointing where 10 should be!

I'm no more able to explain this than I am to account for the sequence in which I unfailingly envisage the months of the year (Figure 5). For all other number sequences apart from the hours of the day an entirely different system prevails as is also shown in Figure 5, and this, peculiar as it is, I can account for. It will be noted that the critical points are at 6, 12 and 20. After 100 the series starts again, and similarly after 200, 300, etc. Yet another series starts with 1,001, so that the number of Don Giovanni's amorous conquests in Spain, for instance, 1,003, seems to be just two places vertically above 1,001.

I haven't the slightest doubt that this otherwise inexplicable pattern comes straight out of the road system of Stibbard as perceived from my bedroom window. The first (vertical) leg takes one as far as the King's Arms (Figure 3) where the road makes a right turn and disappears between high banks out of sight. This turn corresponds with Number 6. Number 12 occurs at the T-junction between The Lodge and the toll-house, which is Number 20. At this point the sequence makes a left turn and follows the turnpike road to Norwich (Number 100).

It will be apparent that the scale is far from consistent. Each 'leg' has its own scale. Thus 1-6 covers less than a quarter of a mile; 6-12 rather more; 12-20 about half that distance, and 20-100 about 20 miles, but such inconsistency is hardly surprising in the concept of a very young child. That I was still very young when this pattern became fixed I have no doubt. Although a dozen and a score were still the important numbers in the English monetary system (pence and shillings) and were regularly employed in counting, for example, eggs and other farm produce, the decimal system was far more basic in all arithmetical calculations, and had I been older when formulating this system it's much more probable that 10 would have occupied a point of major significance. Decades of experience of the decimal system, however, have failed to eradicate this pattern which is how I still visualise number sequences today (except for the hours of the day!).

The roads which formed the basis of this pattern comprised, of course, only a small part of the whole road system of the village.

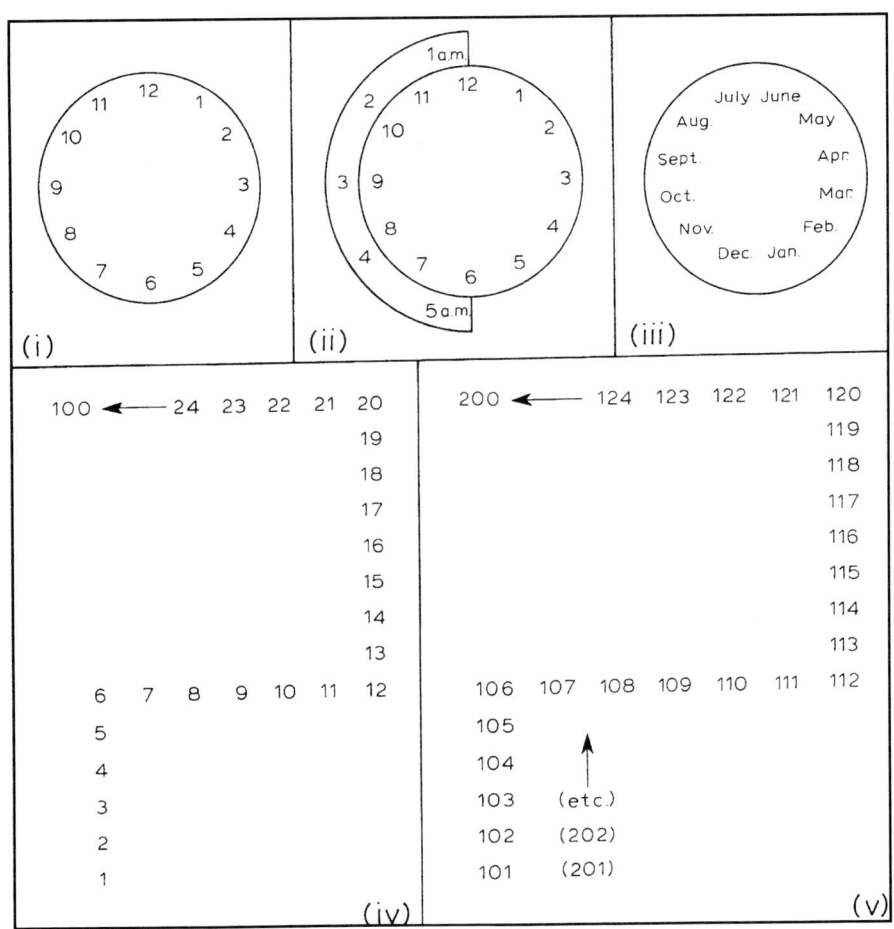

FIGURE 5. THE NUMBERS GAME. Some regular patterns and arrangements in which numbers are perceived. (i) The time; (ii) The time between midnight and 6 a.m.; (iii) The months of the year; (iv) and (v) Numbers other than those indicating time. For further explanation see text. Drawing by Keith Scurr.

FIGURE 6. 'INSEPARABLE AS FATHER, SON AND HOLY GHOST'.

Demonstration of wheat-threshing, as practised in the nineteen thirties, at the Museum of East Anglian Rural Life, Stowmarket, Suffolk, August, 1986. The sheaves, stocks or shocks are pitchforked on to the drum (centre) where the threshing takes place and the straw is removed up the elevator (left) for carting away or building up the straw stack. Photo by the author.

This system became highly familiar as I accompanied my mother on her frequent visits round the parish. It was supplemented by a second-order system of pathways, some trodden regularly, others not. These tended to be more exciting to explore. One was less likely to meet other people, more likely to establish that intimacy with nature which is the special privilege of the country child.

When I was six or seven, I can't remember which, Helen and I began to take lessons with John Ashworth, whom you have already met. His father was the largest farmer in the village and lived at The Lodge. Our governess, Miss Marler, cycled daily four and a half miles each way from Hindolveston to teach us. To reach The Lodge we used to walk every morning along the field path through the meadows, and this became more familiar than the road system. It commenced at a stile near the village green or 'pound', as it was called, though goodness knows when it had last been used for that purpose. The journey was punctuated by a series of episodes like the little bridge with a white painted rail which spanned a tiny ditch, the crossing of the avenue which led to the farm buildings, and so on. All these made a deep impression on me and I was conscious of being strongly attracted by certain objects or arrangements of objects which seemed to possess the power of causing a particular kind of pleasure not felt in any other kind of experience.

Perhaps the most potent of these was a group of tall trees, poplars and willows, I believe, on the northern side of the meadow. They rose out of the low-lying damp ground by the stream, and I think of them as always backed by firm-edged white clouds floating in a blue sky. They became a kind of prototype for a whole class of similar pleasurable experiences. The sense of projection upwards and the awareness of other even more distant objects, the clouds, seemed to trigger off something inside me which I couldn't have explained except perhaps in some primitive theological way. ('There's a home for little children/ Above the bright blue sky'?) I don't suppose I seriously attempted a rational explanation of this experience, but I was deeply aware that it was important to me, and I didn't expect grown-ups to understand. Already I knew my world was different from theirs.

The Lodge was much more than the place where we had our lessons. It was an elegant farmhouse approached by a gravel drive and set in well-kept gardens, but the really exciting part lay behind. This was the farmyard with its associated working buildings. The landscape of gentility stopped along a well-defined line and gave way to a landscape of dung and urine, of huge expanses of mellow red brick with dark cavities penetrated by nesting swallows, by farinaceous odours and by small children with inventive and rapidly maturing imaginations.

The best place was the hay loft, a veritable sea of hay, straw or both. This was before the general use of baling, and it lay there in a great amorphous heap ready to swallow us up. We burrowed in it, we tunnelled through it, we pulled it over ourselves, we inhaled its fragrance and we choked in its dust. There was a large opening at one end and a hoist to raise and lower things, and through it we could peer down from a place of concealment and watch 'the men' as they came in from the yard. On one occasion Tom Ashworth arrived just in time to see us lowering his daughter, John's younger sister, on the hoist. He expressed himself in terms which ensured that the operation would not be repeated.

This was my first introduction to the agricultural industry as opposed to the agricultural landscape which lay all around us. The barn was full of fascinating objects - ploughs, harrows, binders, machines for chopping straw, machines for chopping turnips, machines, as we were warned, for chopping the fingers of the inquisitive and the incautious, as well as numerous pieces of equipment the purposes of which we never knew. At certain times of the year something quite dramatic would happen. At harvest the threshing outfit would arrive - engine, drum and elevator, as inseparable as Father, Son and Holy Ghost (Figure 6). We were usually allowed to climb on the engine, and, if we were lucky enough to find a friendly driver, we might even be allowed to throw coal on the fire through the firehole door. The smell of the grain and the roar and rattle of the drum is an easily acquired taste and it remains with one for life.

There was only one thing to surpass the threshing equipment for excitement and that was the steam plough, because that required *two* traction engines, placed at opposite sides of the field, to draw the plough backwards and forwards between them. Yes, The Lodge could be a very exciting place!

But I've digressed again. I was telling you about the system of roads and pathways which were the channels of exploration of the world around me. Lower still in the hierarchy was a third-order system, more extensive and, on the whole, more exciting. It consisted of various lines of passage, unrecognised as public rights of way, but, in a tolerant and friendly farming community with a general respect for the Established Church (in spite of the 'tithe wars', to which I shall refer later), nobody was going to throw out the parson's kids as long as they respected what later became known as 'the countryside code'.

Respect it we invariably did, at least as it was then understood. We never left gates open or interfered with stock, though we did engage in some practices which would later be frowned on not only by well-bred country-lovers but even by the law. We were never discouraged, for instance, from collecting birds' eggs, provided we

FIGURE 7. 'DITCHES WERE A PARTICULAR SOURCE OF DELIGHT'.
This is the ditch which contained the minuscule iron deposits. It forms part of the
parish boundary between Stibbard (on this side) and Fulmodestone. Note how it is
hedged on one side and open on the other. It was customarily kept more clear of
vegetation than it appears in this 1986 photograph by the author.

FIGURE 8. A TYPICAL 'PIT'. This is the marl-pit where we used to observe newts from the branches of the tree shown in the picture. Note how the margins are alternately open and enclosed with vegetation. The pit is marked on Figure 3 some 200 yards west of The Lodge. Photo by the author, 1986.

took only one from a nest, never left less than three (birds could count up to three, we understood, but not four), and didn't frighten the mother bird away. We were positively encouraged to pick wild flowers as a training exercise, and a very good one, in the techniques of recognition and classification. Hard luck on any rare orchid that might have been among them! The more unusual the species collected, the greater the commendation we received on presenting them to our admiring parents; but it must be remembered that this was in the days before the general use of herbicides, and the meadows still furnished such a profusion of wild flowers that nobody questioned the propriety of a couple of small children helping themselves. Then there were the various foodstuffs provided by nature. These too we collected and duly consumed, though not, of course, those raised by the farmer, which we never touched. Blackberries, crab-apples, mushrooms, watercress and so on, all went into our little baskets, or where suitable for immediate consumption, straight into our mouths.

The system of pathways which gave access to this treasure house was made up of various components. It started in the fields adjacent to the garden and its ramifications spread along the stream courses and their tributary ditches, under the hedges, through little copses and, in fact, by any channels which our exploratory initiative could discover or our ingenuity devise.

Ditches were a particular source of delight (Figure 7). They were usually flanked on one side by a hedgerow and lay open on the other side to the field whose boundary they helped to form, but they were often deep enough to afford concealment, at least to little people, even from that direction. The wellington boot solved the only serious problem which faced the explorer of the watercourse. Only after heavy rain did these little ditches carry more than a trickle of water, but they contained a richness of plant and animal life. We would also occasionally encounter the little circular opening of an unglazed pipe where a field drain discharged its water, and every now and then we would come across deposits of iron-stained silt and orange mud in the streambed. Only we knew exactly where these were, and, provided we kept the secret until we were grown up, we would one day be able to put this little village on the map with the only blast furnace in Norfolk. There must have been enough iron in those rich alluvial deposits to cast at least a couple of manhole covers.

For a more copious provision of water we had to look elsewhere. The little stream which ran through the meadow in front of our house was just big enough to contain real fishes. Exceptionally they probably attained a couple of inches. For deeper water and larger fish we went to the ponds of which there were

many in the surrounding fields. Later I learnt that the glacial soils of
the area varied greatly in their heaviness and that it had been the
practice in bygone days to dig pits in the heavier clay or marl and
spread the material extracted over the sandier fields thereby giving
them more body. All this had happened many years, perhaps
centuries, earlier, and by the time I first saw these marlpits they were
well and truly mellowed. Most of them were fringed with
hawthorns and sometimes larger trees at least on one side (Figure 8),
though they were usually left open at some point to allow easy access
for stock. In a mixed farming countryside like North Norfolk the
ability to provide a supply of water greatly enhanced the commercial
value of any field.

Since these ponds were not fed by surface streams they showed
no inclination to silt up, and, though I never remember one being
dredged out, many of them were probably almost as deep as when
they had first been dug. My wise mother judged it better to rely on
our realistic understanding of the danger of being drowned than on
any general order of prohibition. We were frequently admonished to
be careful but rarely if ever to stay away.

The ponds, or 'pits', as they were called in the vernacular, were
very exciting. The water was usually muddy enough to obscure the
bottom except near the very edge, though shafts of sunlight could
penetrate to a certain depth, and, with a little help from the
imagination, disclose all sorts of improbable shapes. Very
occasionally these pits harboured fish, but there was a much better
chance of finding newts, and I have vivid recollections of these little
dragon-like creatures floating with fingers and toes extended just
below the surface and every now and then popping up for a gasp of
air and making a little ring of concentric ripples. Lesser creatures
like tadpoles proliferated in their proper season.

The visual images which I retained from these exploratory
excursions no doubt presented themselves as self-contained pictures
needing no wider context to make them meaningful. Nevertheless
their geographical occurrences were keenly noted. Every visual
experience was somehow mentally recorded, perhaps subconsciously
and not always accurately, within my emerging mental map. This
was already extremely important to me at a very early age, and,
because it was important, I acquired an above-average capacity to
develop it. Most children acquire a particular knowledge and
understanding of those things which most interest them, whether it
be food or flowers, sporting heroes or postage stamps, and for me a
fascination with the geographical relationships between places
matured far earlier than most of my other interests and aptitudes.

I can see now, in retrospect, that my success in building mental
maps stimulated my thirst for more environmental information, that

is to say it drove me into more and more exploratory adventures, while the fruits of these explorations furnished me in turn with more and more raw material for my mental maps. I think a contributory cause of this may well be that my schoolteacher mother introduced me early to the technique of making maps. She taught me, in short, to convert images seen more or less horizontally from particular but randomly distributed viewpoints into areal patterns on horizontal plane surfaces. The result was that skills in understanding maps and diagrams, which I have sometimes experienced difficulty in communicating to undergraduates, I had myself mastered as far back in childhood as I can remember.

Unfortunately (or perhaps not!), this precocity extended to no other areas of my education. Had my mother been able to arouse an equal fascination in the magic of numbers I might not now find myself having problems in mastering concepts which seem to be crystal clear to quite small children in this computerised world. 'It takes all sorts to make a world', we say, and we might add that each sort makes a very different sort of world. The world I made was neither better nor worse than anybody else's; it was, however, different.

Once the principle of fitting environmental experiences into a comprehensive geographical framework was established it could be applied over a more extensive field. Neighbouring villages gradually became familiar. Visits to Fakenham, four miles away, were undertaken to supply such wants as could not be satisfied at one or other of the two shops in the village, Miss Boulter's or the Post Office. Visits to Norwich, twenty miles away, were less frequent, but, no doubt because of their comparative rarity and greater distance, certainly no less memorable. Often, it's true, they were memorable in part for their purpose, if this happened to be to go to a pantomime, for example, but in part also for the experience of getting there and back.

By venturing farther afield we were able to discover places which afforded opportunities for doing all the exciting things we had done on the home ground, only now on a larger scale. The limitations of the village stream and the scattered ponds were dramatically shown up by the Wensum which contained fish big enough to eat! One could watch them from Great Ryburgh bridge (Figure 9) slinking among the trailing green tresses of water weed. The shrubberies of the garden and even the little copses in the outer parts of the parish seemed to be mere appetite-whetters for the deeper glooms of the Severals. Once a year we used to go to the even deeper glooms of the woods at Swanton Novers, four miles away to the north-east, to pick lilies of the valley. These emitted one of the few flower smells I came to recognise as a child, but the pleasure of smelling them had to be postponed until we were home again, because the act of picking could only be attempted after a most liberal application of fly-repellent so pungent that no other smell could compete with it.

This was my most potent experience of 'the forest' and it furnished food for my imagination to work on when I was introduced to the North European forest through the fairy tales of the Brothers Grimm, Hans Andersen and lesser story tellers. The fact that Swanton Novers Wood was quite unlike the coniferous forests of Germany and Scandinavia was not important. If you want to know where Hansel and Gretel lived, it was, I assure you, at Swanton Novers, and although I never found any witches or gingerbread houses, it wasn't for want of looking.

The other principal source of attraction in North Norfolk was the sea, and we used to go as often as we could to the beaches from Holkham to Sheringham and occasionally further afield still. Our nearest, and, I suspect, our favourite place was Wells-next-the-Sea, a misnomer if ever there was one, because the little town lay near the end of a channel a mile or so back from the beach, while at low tide the sea retreated yet another mile towards the north. The broad expanse of exposed sand, however, was full of cockles which I enjoyed helping to collect, though nothing would have induced me to eat one, and, more to the point, it was laced with smaller tributaries by which the receding waters found their way to the main channel and hence to the open sea. Within these lesser channels tiny shrimps tickled the bare feet and baby flatfish flipped and flopped along the sandy bottom.

It was at Wells that I first practically participated in the science of military architecture, and perhaps it was those hours of inventive sand-castle design and construction that, several decades later, induced me to draw on the imagery of the castle as the best metaphor for conveying to you my feelings about my childhood home. It was at Wells, too, that I learnt the basis of another science, that of the water engineer. Little channels were dammed by weirs; new watercourses were constructed in such a way that, when the tide came in, they would undermine the castles. At Wells I found out not only how to observe the environment but also how to manipulate it.

Wells, like its neighbour Holkham, had earlier been the scene of some interesting experiments in the reclamation of the salt marshes and the stabilisation of the sand dunes by planting them with conifers (Figure 10). This provided an unusual variation on the woodland theme. Because no ground vegetation other than the occasional tuft of marram grass would grow in the pure sand under the pine trees, games of environmental perception were reduced to extremely simple terms. The tree trunks were just about wide

FIGURE 9. RIVER WENSUM FROM GREAT RYBURGH BRIDGE, NORFOLK.
View upstream to the north-west. Photo by the author, 1986.

FIGURE 10. ABRAHAM'S BOSOM, WELLS-NEXT-THE-SEA, NORFOLK.
Boating lake with pine plantations on sand dunes behind. View to the north-north-west. Photo by the author, 1986.

enough for people of our size, though not a grown-up, to hide behind, but between them there lay long narrow shafts of open space. Hide-and-seek was somehow different at Wells-next-the-Sea.

Farther east the coast is one of the most interesting areas of Britain for geomorphologists and bird-watchers, but for us it was virtually inaccessible, being separated from the undulating farmlands by salt marshes with extensive creeks. By the time the beach became accessible again little feet had to cope with interminable banks of shingle to reach the alluring sandscape with its deep pools gouged out by the retreating waters swirling round massive timber groynes. Cliffs began to compensate, it's true, but for me the really memorable thing about Sheringham and Cromer was nothing to do with the beach or the sea; it was the landscape of sandy hills a mile or so inland. This gave rise to another type of woodland, different from that of Wells and Holkham, a mixture of silver birch and Scots pine with extensive openings of heathland here and there commanding a view into the woods and in places over them and out to sea. It was my first acquaintance with an association of soils and vegetation which has become established in the English public mind as 'recreational land' *par excellence*. A more familiar variant of it occurs in the Surrey commons. In fact the name of this particular spot behind Sheringham is Pretty Corner, and on this point at least I never felt any inclination to diverge from the commonly accepted public taste which its name implied. It didn't occur to me to ask what made it pretty, at least not for many years. When I did eventually find myself asking questions of that kind, Pretty Corner, or rather the type of landscape it represented (it has, alas, been somewhat altered since), figured prominently in the argument.

As I ventured forth to these places it was not simply the destinations which interested me. The roads themselves by which we travelled there and back became familiar, and their various sections were fitted into my developing mental map. Gradually I was able to anticipate what I was going to see round the next corner. I became aware also of strong preferences for particular stretches. Some were comfortable, others less so. I greatly enjoyed these experiences of serial vision, so it was perhaps not surprising that I should relish the opportunities, when they came, to practise this on a larger scale. The ultimate experience of this kind was the journey to Bradley Hall. I promised to tell you more about Bradley Hall, and so I shall; but following Stevenson's maxim that to travel hopefully is a better thing than to arrive, let's begin our review of the wider world in the next chapter by following the Bradley road.

2 FURTHER STEPS IN EXPLORING

'Methinks 'twound heighten joy, to overleap
At will the crystal battlements, and peep
Into some other region . . .'

In the intervals between his regular attacks of malaria my father's health had taken a turn for the better in the rural atmosphere of Stibbard, and we hadn't been there long before it became apparent that the work involved in looking after a parish of 373 inhabitants was not going to use up all his energies. Rather than seek a larger parish, however, he decided to fill the gap by taking on some of the administrative work of the diocese. This provided us, I dare say, with more opportunities to visit Norwich than might otherwise have come our way, but in the main it involved correspondence which my father could conduct from his own study, and the work was less susceptible to serious interruption by illness than would have been the work of a larger parish, since he could do much of it in his own time and at his own pace. Provided, therefore, that he could make satisfactory arrangements to cover Sunday services and other essential parish obligations, there was only one obstacle to our being away for three or four weeks every summer. That obstacle was finance, because, although we were more affluent than most of our immediate neighbours, there was not much to spare after regular outgoings had been provided for.

All this led to the establishment of a regular pattern of summer holidays. With the exception of one or two special occasions, for which I suspect my grandmother made financial provision, hotels and boarding houses were out. The only place we could afford to go to for any length of time was my grandmother's house. So Bradley became a kind of second home base, contributing wholly different ingredients to our emerging concepts of the world, and our annual visit was anticipated with great excitement. Once or twice we made the journey by train, but this was unusual. Our earliest form of private transport was the donkey and trap. Needless to say we never attempted the two-hundred mile journey to Bradley in that! It was only a short time, though, before the demise of Natty Bell pressured my father into acquiring a more ambitious vehicle (or perhaps it was the other way round!). This was an A.J.S. motor-cycle combination, and it made a number of journeys across the breadth of England as well as to places nearer at hand. Although we only had it for a few years, they were among the most formative years of my life, up to the age of eight, and it therefore achieves a disproportionate importance in my story.

When I recollect, from later experience, the problem of going away with small children, I can't imagine how my mother contrived to pack all our requirements for a month into the available space. Since she was riding pillion, the only place for luggage was in the back of the side-car, the front of which was occupied by my sister and me, sitting side-by-side. This was our travelling order as we set out for Bradley.

On one occasion, round about my eighth birthday, we timed our annual visit so that we should arrive before one of my father's malarial attacks was expected. Unfortunately we mistimed it, and my grandmother was convinced that the rigours of the journey were a contributory factor in causing my father's condition. This is how we came to find ourselves presented with a four-seater twelve horse-power Clyno, and our later trans-Pennine journeys were undertaken in the relative luxury of what today would be called a 'convertible' but in those days was known as a 'tourer'. It meant you could take down the roof, or what was known in England as the 'hood' (to the confusion, no doubt, of Americans), when the weather invited a breezy drive.

The Bradley Road

I will not try your patience with a mile-by-mile description of our route. My purpose is rather to draw attention to the particular way in which my little brain tackled the task of informing itself about the changing environment as we passed through it. We all notice only some of the things we see and remember only some of the things we notice, and it is this process of selection which enables each of us to build a different world out of common ingredients.

As the route became more familiar through repetition I came to anticipate particular features remembered from previous occasions. They were of many different kinds. Some were trivia, highlighted for our amusement by parental comments made to help the miles along. So we always looked for the sign near Holbeach which read 'Family Howling Butcher' and I remember how dismayed we were when it was changed to 'Howling & Son, Family Butcher' or words to that effect. It was as though the whole Fenland had suddenly lost its sense of humour, a redeeming feature of which it stood sorely in need, since it was the only place in the whole journey which I found not only dull, not only alien, but almost hostile.

I need now to remind myself of this attitude, because in later years I have found this area a source of great fascination, but it's a fascination born largely of intellectual understanding rather than of spontaneous emotional response. The story of its transformation over the centuries from the seasonally waterlogged waste of Hereward the Wake into one of the most fertile plains of England can be monitored in a thousand minutiae hidden within the landscape, but they have meaning only for those who have cracked the code, and in those days they escaped even perception let alone interpretation.

So our first view of the Fenlands as we came down the gentle hill into Kings Lynn was of a flat plain of spaciousness stretching away beyond the town apparently for ever. It required an act of faith to believe one would ever emerge at the other end, and by the time we crossed the Great Ouse at the Freebridge and were virtually at sea level, even that distant panorama had disappeared, the horizon had been foreshortened and one felt a sense of being swallowed up in a huge expanse of nothing in particular.

Of course, there were really plenty of things to look at. We passed the proliferation of splendid medieval churches where early settlement had been possible along the slightly higher land of the silt belt, and a few miles to the north we could see the magnificent fifteenth-century 'Stump', the tower of the church of Saint Botolph which had given its name to Botolphstown, abbreviated to Boston, and thence by process of colonial transfer to its young namesake in Massachusetts. But these were all, so to speak, punctuation marks in what could only be described as a tedious experience in comparison with the more undulating parts of the road. Remember, in the motor-cycle combination we had a flat-out speed of, perhaps, thirty-five miles an hour and a cruising speed well below thirty, and those forty-five miles from Lynn to Sleaford, roughly a quarter of the whole journey, seemed to take an awfully long time.

Just before Sleaford the first gentle swellings in the ground proclaimed that the ordeal of the crossing of the Fens was over, but still there wasn't much of interest unless we were lucky enough to see a couple of aeroplanes at the R.A.F. College at Cranwell, and we always looked out for a splendid windmill at Heckington. More often than not it was working. Then came one of those really dramatic moments as a huge panorama suddenly opened out and there below us lay the Vale of Trent. Not until I was nineteen was I to

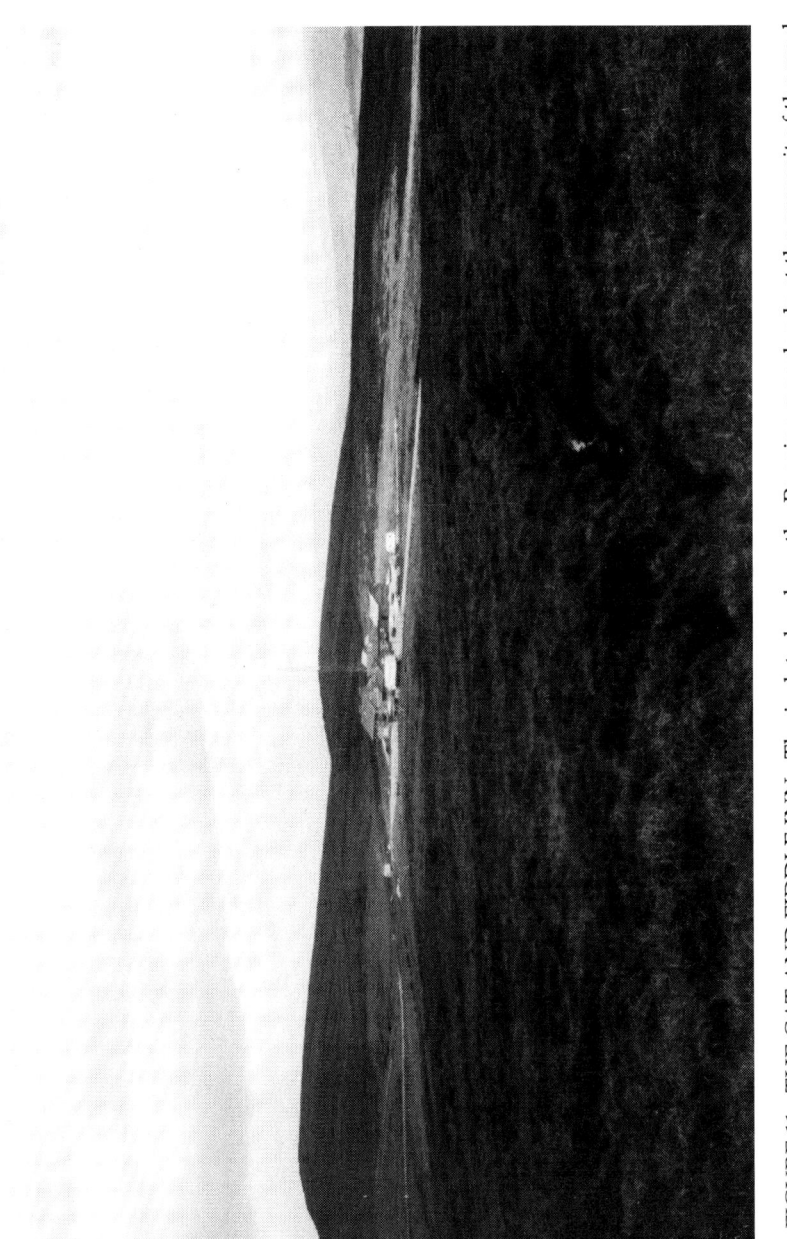

FIGURE 11. THE CAT AND FIDDLE INN. The isolated pub on the Pennine moorlands at the summit of the road between Buxton (Derbyshire) and Macclesfield (Cheshire). Photo by the author, 1987.

FIGURE 12. BRADLEY HALL, LANCASHIRE. View from the south-west. In this nineteen thirties photograph note the large conservatory on the south side. The oakroom window is on the ground floor of the bay (left) and my grandmother's bedroom is above it.

discover that we were here on the crest of the escarpment of the Lincolnshire Limestone. The climb up the dipslope had been so imperceptibly gentle that the revelation when it came never failed to fill me with excitement. The Lincolnshire Limestone escarpment is in fact a double one, and half way down its face we passed through the village of Leadenham sitting comfortably on a little shelf beyond which the road completed its descent into the valley. I used to think how fortunate this little village was in comparison with those bleak, exposed villages in the Fens. Sheltered from the east by the upper escarpment it nevertheless was still high enough to command a broad view over the western plain. Lucky, lucky Leadenham!

Our interest in the various buildings we passed was enhanced by parental information, probably the response to bombardments by questions shrilly squeaked out from the side-car in the teeth of the wind. Newark Castle and Kelham Theological College were landmarks, as was the crossing of the 'smug and silver Trent'. So on through the sandstone country around Mansfield and the first view of a coal-mine. Long before I knew anything about geology, or that we were here crossing successive outcrops of strongly contrasting rock types, I was aware that the whole nature of the country changed rapidly, and when we encountered particular objects, like the crooked spire of Chesterfield, these were not isolated phenomena recollected out of context, but units which had their proper places in a continuum of experience a couple of hundred miles in length.

Beyond Chesterfield and Baslow the road followed the valley of the Derbyshire Wye, and in the immediate approaches to Buxton this afforded a quite different type of visual experience. I was later to encounter hundreds of examples of narrow wooded valleys, many of them awarded the designation of beauty spots, but this was the first that became familiar to me. Neither was I convinced by my mother's protest that the valley had been ruined by the Buxton Sewage Works. That too had a place in the appetite of the enquiring mind.

I was by no means dismissive of the architectural splendours of Buxton, still exuding the genteel air of a Victorian Spa, but excitement was already mounting in anticipation of the climax of the whole journey, the ascent of the Cat and Fiddle, where the road crosses the main watershed of England beside the isolated inn of that name which stands on the very summit at nearly seventeen hundred feet above sea level (Figure 11). The approach took us through yet another highly distinctive type of landscape, the open moorland of the Peak District. It would be difficult to conceive of anything in England more different from the gently undulating hedged fields of Stibbard. As for the moment of arrival at the summit, if the view of the Vale of Trent from Leadenham surpassed that of the Fens from the Fakenham-Lynn road, the view out over the Cheshire Plain was surely supreme!

The plain itself was something of an anti-climax. It was pleasant enough and my mother greatly admired the neatness of its trimmed hedgerows and the smooth grass of its lush pastures. But to me, who had just descended from the giddy heights, it looked rather like a landscape trying to be Norfolk, and not too successfully.

I knew that there was one more radical change still to come as we crossed the Mersey and the Manchester Ship Canal into Warrington and Industrial South Lancashire. The last few miles of our journey took us, in fact, through the South Lancashire Coalfield. This was still some years before George Orwell wrote *The Road to Wigan Pier*. It was like entering a foreign country. Not only did the people speak a highly distinctive variant of the English language; they wore native dress which contrasted almost unbelievably with what I was accustomed to in rural Norfolk. Women of all ages, not just the elderly, wore shawls over their heads, and many people, men, women and children, clattered along in wooden clogs. Some wore no shoes at all for reasons which will be obvious to you even if they had to be spelt out to me. Already before the economic recession of the 'thirties the road into Wigan from the south through Poolstock must have been as expressive of urban industrial poverty as anything in the England of Dickens. It would have been difficult to devise a two-hundred mile route that could have provided a more complete catalogue of landscape types epitomising the face of England.

The coalfield continues for a few miles north of Wigan, but the more rural matrix into which the collieries hereabouts were set came as something of a relief after the concentrated grot of South Wigan. As we turned down Bradley Lane from the main road at Standish we still had one active coal mine to pass at Broomfield, with its bulky spoil heap towering above us, but it was set among green, or at least grey-green fields, and it commanded a wide view eastward towards the Pennines. It was a kind of half-in-half landscape, a shot-gun marriage between rural and industrial if not urban England, and it was out of this landscape that we turned into the little oasis of rhododendrons flanking the orange gravel drive to be confronted by the white-windowed, heavily-gabled object of our quest, Bradley Hall (Figure 12).

Bradley Hall

If the analogy of military architecture was appropriate for my description of Stibbard Rectory it would, I'm sure, be no less fitting for Bradley Hall. In Stibbard I could at one time have put a name to every single inhabitant of the village. The natives could be presumed to be friendly and many of them were indeed my personal friends,

FIGURE 13. THE OAKROOM, BRADLEY HALL. The room, as it appeared in the nineteen thirties, is uncharacteristically well lit for the purpose of taking this photograph. I remember it as a dark place made darker still by so much oak.

whereas the population which inhabited the territory surrounding Bradley was wholly anonymous, and anonymity is the first step towards potential hostility. They were, I was assured by my grandmother, the salt of the earth, but I didn't *know* them, so the strategic symbolism of the defended place would be even more in order. Never mind; I'll spare you a second dose of keeps and baileys. You can fit the image to the terminology if you so please.

The house itself was of many periods. The earliest parts were thought to be medieval and it had been modified at various dates, not least in the nineteenth century. The interior I remember as a rather gloomy place, though you must understand that this was in no sense a pejorative word. It is perhaps even arguable that my experiences of Bradley were a contributory cause of my being more tolerant of gloomy places than most people seem to be. It contained a good deal of oak, so dark that it was almost black. The whole ensemble made an unforgettable impact and not only on the visual sense. Half a century later my sister and I have simultaneously exclaimed 'Bradley cupboard smell!' when confronted by an odour for which no other term was either available or necessary. I have always imagined, admittedly without evidence, that it was produced by drips of alcohol trickling for decades down the outsides of bottles and chemically interacting with seasoned English oak.

The drawing room was called the oak room (Figure 13) and the adjacent room the gun room. The latter was approached by a short, dark passage in which the grown-ups used to hang their coats, purposefully, so it seemed, to provide a screen for us to hide behind. The ultimate in dark, gloomy places was the priest's hiding hole which was alleged to lie behind one of the bedroom fireplaces. There was no way we could get into it, if indeed it existed, and the intention of one of my uncles to open it up was thwarted by the onset of the Second World War when the whole place was requisitioned by the Ministry of Defence. By contrast almost the entire south side of the house was covered by a large conservatory which would have been even lighter and brighter had it not been filled with ferns, for which my grandmother had a great liking, and other foliage. Like the oak cupboard it also had a highly distinctive though quite different smell. It contained a little brick-lined pit with mossy edges filled with water to maintain the humidity of the atmosphere. It was probably the first place to confront me with that very important question in landscape aesthetics, 'When does indoors stop being indoors and start being outdoors?'

It was, however, the garden that had the most influential impact on my developing taste. The house was surrounded on almost three sides by a lawn, part of which was at times in use as a tennis court. It contained a rather fine weeping ash which wept so effectively that when we were under its drooping branches only our feet were visible. But there were much more effective hiding-places, much deeper glooms, in the rhododendron bushes which fringed the lawn and continued as a kind of extension down both sides of the drive. Although I didn't realise it until much later, I was exposed at Bradley to prolonged doses of intense radiation by the very essence of middle-class Victorian (or perhaps Edwardian) England, an experience which was to fortify me later in what seemed to be a lone resistance against the cynical and contemptuous attitude of my generation to a cultural expression which it generally despised.

The rhododendron bushes afforded a totally opaque screen all the way round the outside of the lawn. Being evergreen even in the winter they were as impervious to the prying eye as the brick wall at Stibbard, but, unlike the brick wall, they could themselves be penetrated by our small persons. There, under the dense canopy of glossy green leaves, was another world, a kind of perpetual penumbra through which we could thread our way between the branches from one end of the shrubbery to the other, totally undetected by eye or even ear, unless we recklessly stepped on a dead twig.

Every now and then we encountered exciting things like little wells. Every now and then also the linear passage opened out into little chambers of twilight. Many of them had names, mostly bestowed by my cousin Charles who lived with his brother John and my Uncle Henry and Aunt Phyllis in Bradley Oaks, a more modern house adjacent to the entrance to the drive. 'This is Pussy's House', he would say of one such little chamber, and of a shallow depression in the ground 'this is Pussy's Sick-basin'.

Being in a coalfield with much industrial activity going on in the vicinity and long before Clean Air Acts were thought of the stems and branches of the rhododendrons attracted a good deal of grime, and I remember my long-suffering mother trying to clean me up and asking whether it was really necessary to spend so much time in there when the sun was shining so beautifully outside. I urged her to come inside so that I could share with her this world of my own discovery, but her reply was that grown-ups like different things, and that one day I too would grow out of it. She wasn't wrong about many things, but it hasn't happened yet.

On the opposite (eastern) side of the drive the screen of rhododendrons was backed by a strip of more open space separating the bushes from the boundary fence of the garden, beyond which lay the meadow. It was still covered by a canopy of trees but there was more open space at ground level. Here one of my uncles erected a railway, using old colliery rails and a trolley which he rebuilt with a massive oak frame fitted with wooden seats. It was propelled by a good push down a steep incline at the bottom of which it had acquired enough momentum to carry it for twenty or thirty yards along the level. Since it had no means of arrest and must have weighed several hundredweight it was indescribably dangerous but compellingly fascinating, and for some reason I shall never understand none of us ever lost a limb or suffered more than superficial bruises.

Of all the things at Bradley the one which made the deepest and most enduring impression on me was the view to the east across the adjacent meadow. On the distant skyline was Rivington Pike a kind of bastion projecting towards me from the main axis of the Pennines and reaching an elevation of fifteen hundred feet. In front of it a series of gently rolling hills sloped down to the River Douglas which, however, was out of sight because between me and it there lay the embankment carrying the main line of railway from London to Carlisle and Scotland. Entrance to the meadow was obtained through an iron gate set in a wooden frame on the top of which I could sit dangling my legs and taking in the view. Year after year I used to sit there. I had a special dispensation to leave the table early after meals in case I should miss this train or that. In those days this was quite a concession and I suspect it was procured through the advocacy of my lawyer uncle, Uncle Edge, at that time a bachelor living with my grandmother. I shall have more to say about trains later in the context of obsessive interests. Here it will suffice to note that this is where it all began.

One other memory of Bradley is worth a mention. The window on the stairs, which faced south, contained a certain amount of stained glass, and one could therefore look at the same landscape through different colours, rather as in a Claude Glass only with several colour options. In the foreground my grandmother's rose garden sloped gently down to a small stream which flowed under the railway to join the Douglas. Beyond the stream lay some woods of mixed deciduous species and beyond them, rather over a mile beyond, though apparently rising out of their leafy canopy, there sprouted the chimney of the Victoria Colliery. It was well named, being the embodiment of the spirit of Victorian industrial architecture, grimy, phallic and, to the eyes of most of my relations, monstrously intrusive; but I came to love it dearly, and not only it but the whole species of which it was the type example.

From that window also I learnt another lesson, that the landscape as felt contains more than the landscape as seen. The view from my bedroom window at Stibbard had told me this, but here the lesson was driven home. Beyond the chimney the land surface sank into a kind of saucer through which flowed the River Douglas and in that saucer lay Wigan. One could see nothing of it except the occasional plume of smoke from its industrial premises because the wood was in the way; but it could be sensed, much as one can sense somebody in the next room when one knows they are there. Everything I came to know about Wigan, my uncles' office, the dentist's, the locks on the canal - all things visible and invisible - wormed their way into the view from that window and became fused together in a single emotional experience.

By the application of the same principle I felt the same way about the Pennines. Rivington Pike was a sort of symbol of what lay behind, and I couldn't look out on that eastern horizon without 'feeling' the main body of the Pennines themselves as I had seen them further south at the Cat and Fiddle. What I could see with the eye was only the beginning of something which stretched away beyond the field of vision, beyond Buxton and Chesterfield all the way to Stibbard. It was as though one could 'know' the whole breadth of England in a moment of time, as though one could compress sequential experiences into one synoptic experience, as though there was some relationship between time and space which one could dimly sense but not explain. One day, perhaps, I would think about such things more seriously.

Beyond Bradley

During our visits to Bradley we invariably made an excursion to see the Mathers, my Uncle Edward, Aunt Gladys (my father's surviving sister) and Cousins Charlie, Helen, (known then as Rose) and Alfred. Uncle Edward was another clergyman and his garden abutted on to the grounds of Hornby Castle which contained peacocks as well as the inevitable rhododendrons, and within them we were frequent trespassers. Our cousins were slightly older than we were. We admired them greatly and were always ready to learn from them. They introduced us to the country round about, which, although still in Lancashire, lay well to the north of the coalfield and could hardly have been more different. It was an area of pastoral farming with dry stone walls as well as hedges, and both the Lune and the Wenning, which joined there, were very different from the sluggish streams of East Anglia, where water surfaces were smooth and reflective and rapids almost non-existent. These Lancashire rivers, by contrast, were just emerging from the Pennines and their watercourses contained boulders, pebbles and slabs of rock. Broken water flowed swiftly over and round them, finding a temporary tranquillity in deep pools which separated the rapids. In these pools one could often see large fish, trout and even salmon, rather than the coarse fish of the Wensum. At Hornby, too, we first came to know the birds of the moorland edge, curlew and lapwing, and learnt that different kinds of bird regularly associate with different kinds of landscape. I could not have guessed how important the concept of 'habitat' was to become in my later thinking.

Once or twice during these years my grandmother made arrangements for us all to go to the Lake District for a few days. There I encountered an altogether more dramatic landscape (Figure 14). I believe I must have been eight when I first climbed Skiddaw, maybe seven. Helvellyn came a year later. From these summits my father pointed out the principal peaks. I became quite familiar with them and could recite their elevations to the nearest foot; but more importantly I formed from the map a clear picture of their positions in relation to each other and to the various lakes, and I was at pains to work out these relationships as they would appear when viewed from different vantage-points. The results were probably inaccurate, but the important thing is that I was motivated to try.

I suppose one might interpret this as revealing the rudiments of a scientific way of thinking about the mountains, but, if so, it certainly didn't displace the highly romantic responses they spontaneously evoked. I well remember looking up to Helvellyn from Patterdale at the end of one of those rainy days when the clouds were beginning to break up just as the light was fading. Fragments of cloud and mountain seemed to be inextricably mixed up in the sky; the sense of height and distance was confused and this whole mixture of land and sky was permeated by a feeling of awe. Part of me wanted to reduce the mountains to the stature of measurable objects, to the actuality of mappable information; part of me resented the very idea and cherished them as perpetually elusive mysteries. I found myself flirting with concepts of animism, recognising the mountains as inert masses of rock yet wanting to think of them as possessed of individual personality, like the curiously shaped pair of trees on the Norwich turnpike. This again is a subject we shall return to later.

None of these excursions from Bradley, however, can compare with the one we made in 1926. I was six and a half and we spent a week in the Isle of Man. It was certainly the most exciting experience I had enjoyed up to that time and it provided me with a veritable gallery of vivid pictures which I have retained in my memory to this day. Two episodes have, in retrospect, acquired a particular significance because they furnished information which lent itself to scientific explanation but only after it had been pigeon-holed for many years. The first was the extraordinary clarity of the air and the freshness of the cold wind which whitened the crests of the little waves as we sailed from Liverpool down the Mersey. Distant objects appeared like miniature cut-outs with stark outlines, and as we moved out into the Irish sea the horizon seemed as sharp as a knife-edge yet also an immensely long way away.

The second of these episodes occurred when we went on the little narrow-gauge train from Castletown to Port Erin at the southwestern extremity of the island. The trains themselves were memorable enough with their tall-chimneyed engines. I remember, too, the little octopus in the aquarium at Port Erin and the conflicting feelings of fascination and aversion which it aroused; but the clearest picture I retained in the mind's eye was of the general 'feeling' of this southern part of the island. I don't suppose I attempted to rationalise it or to find descriptive words for it any more than one would try to describe verbally the personality of a newly introduced acquaintance. I just felt that in some inexplicable way something had happened to the landscape to make it different, something had changed its mood, just before it reached the end of its existence and plunged into the sea.

There were, however, many other memories retained from that holiday. There was the driving rain which kept us indoors all day, in compensation for which we were taken after our normal bedtime to Derby Haven to see the eerie light of the sunset which followed the storm, curiously illuminating the huge, thumping waves of a sea still boiling with anger. Although I have seen bigger and angrier seas many a time since, and not always from *terra firma*, none has remained sharper in my memory. Then there was the agonising sensation of running along wet sand the next morning and stepping barefoot on a sharp object which proved to be the tip of a large and beautiful shell presumably washed up and partially buried by the storm. I still have it as a souvenir, and I am looking at it as I write.

But I suppose it's characteristic of the perception of a six-year-old that, of all the memories of this, my first excursion across the sea, the image I retain with the greatest clarity has nothing immediately to do with landscape. It's that of the Bovril Girl, so called from the colour and texture of the breakfast which she threw up and deposited over the rails into the Irish Sea a couple of miles off Douglas. Male chauvinism begins early, and my feelings towards her were not of pity but of anger. Among small children there is a psychological sense in which vomiting is an infectious disease, and I had been doing pretty well in a moderate sea up to that moment. I have forgiven her now, of course, and I suppose in a way it's a sort of compliment to be remembered in a rather special way and even by a rather special name for nearly seventy years!

As my father's brothers and his surviving sister all lived in Lancashire we used to see them on our annual visits to Bradley. My mother's five surviving brothers and two sisters were more widely scattered, but we did on occasions visit them and this gave me a superficial acquaintance with certain other parts of the country also. For our present purposes what matters is not so much who they were

FIGURE 14. VIEW OVER DERWENTWATER TO THE NORTH-EAST FROM BRANDELHOW, CUMBRIA. Note how the open horizons of Saddleback (centre-right) and the slopes of Skiddaw (left) contrast with the enclosed woodland, principally birch, larch and pine, sloping down to the lake. Photo by the author, 1977.

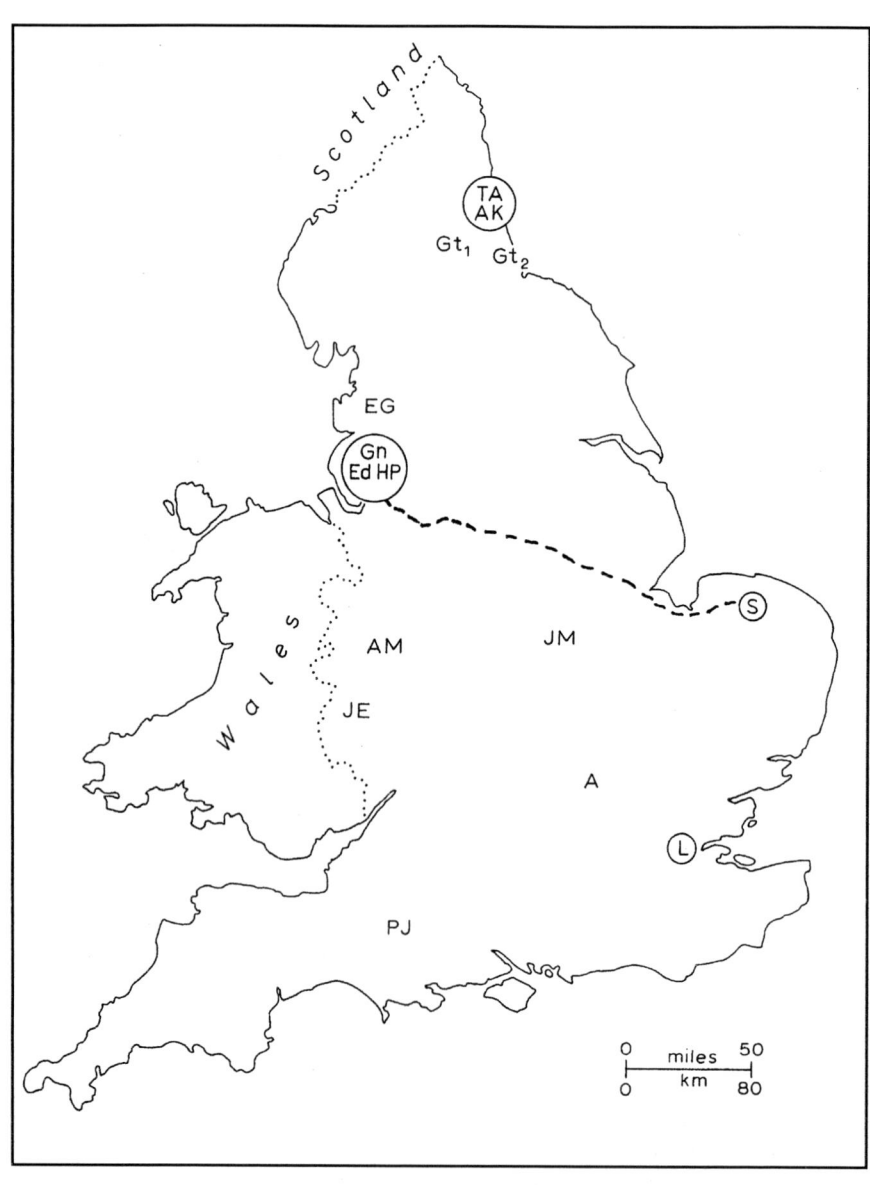

FIGURE 15. 'AVUNCULOGRAM' circa 1925-30.
For explanation see text as well as key. Drawing by Keith Scurr.

Key to Figure 15

Symbol	Relations	Cousins	Surname	Residence
Gn	**Granny**		Appleton	Bradley Hall
Ed	**U. Edge**		Appleton	Bradley Hall
HP	**U.Henry** A. Phyllis	Charles John	Appleton	Bradley Oaks
EG	U. Edward **A. Gladys**	Charles/Charley Helen/Rose Alfred	Mather	Hornby
Gt	**Granty**		Ayers	(1)Westgate (2)W.Hartlepool
TA	**U.Ted** A.Annie	Christopher	Ayers	Gosforth
AM	*****U.Alan** A. Margaret		Ayers	Abdon-under-Clee
JE	**U.Jim** A.Elsie		Ayers	Pembridge
JM	**U.Jack** A.Mollie		Ayers	Ashley
AK	**U.Arthur** A.Kitty	Jimmy Harold Ruth Dorothy Alan Eleanor	Ayers	Newcastle
PJ	U.Percy **A.Jessie**	John Paul David	Opperman	East Orchard
A	**A.Amy**		Ayers	Marsworth

U = Uncle; A = Aunt (blood relations in bold).
S = Stibbard. L = London. - - - The Bradley Road.
*Later re-married Frances (Auntie Fra); Son = Cousin David.

FIGURE 16. THE NEW INN, PEMBRIDGE, HEREFORDSHIRE. My first introduction to the black-and-white half-timbered buildings typical of much of the Welsh Borderlands. Photo by the author, 1986.

or what they were like but where they lived, so, rather than confront you with a conventional genealogical table or family tree, let me introduce you to the 'avunculogram' (Figure 15), probably my only original contribution to the cartographic art.

Uncle Ted and Uncle Arthur, my mother's oldest and youngest brothers respectively, still lived in the north-east, as also did my maternal grandmother, Granty. When we first visited her she lived in Upper Weardale, County Durham, a stock-farming valley in the Northern Pennines. I remember arriving in the side-car late one summer evening and thinking how strange to be greeted by someone I had come to know well from her previous visits to Stibbard and who now seemed to be a part of a wholly alien environment. It was, if one may so put it without impoliteness, like discovering a giraffe in the Arctic; she was just in the wrong place. As a matter of fact she was very much a North-easterner, and when she later came to live with us she never seemed to work the north-east out of her system.

Soon after that visit she left Weardale and moved down to the coast at West Hartlepool, as it was then called, where she lived in an attic flat. I now believe that the view from her living-room window was one of the most influential ingredients in my evolving system of interests and preferences. Far down below lay the main road between West Hartlepool and Old Hartlepool. Beyond that lay the railway, beyond that the docks and the huge sea-water ponds for the storage of timber and beyond that the sea. Ships, cranes and locomotives seemed to be everywhere. It was there that I stayed with my sister on what I believe must have been one of the first occasions on which we were away from our parents. A highlight of the holiday was the day when Granty took us to Middlesbrough on the bus and across the Transporter Bridge, and it was then that I had my first close-up view of an iron and steel works. I returned to Stibbard determined to be an engineer.

Two of my mother's brothers actually were engineers. My Uncle Ted was one and his son, Chris, was later to become one. Uncle Ted and Auntie Annie lived in Gosforth on the northern side of Newcastle. I remember their house for a very particular reason. I went out one day to post a letter for my mother in the pillar box a few yards down the road. I was soon back, running at full speed up the garden path, through the front door and up the stairs into my bedroom. During my brief absence everything had changed. The bed, even, was on the other side of the room! It soon dawned on me what had happened, but not before the lady of the house had followed me upstairs to discover the identity of the intruder. At home, of course, it would have been quite impossible to confuse my own house with somebody else's, but in the elegant suburb of Gosforth it was not difficult. Fortunately my short-term hostess was a very understanding lady, and, since she knew my aunt, matters were soon sorted out, probably over a cup of tea, to my embarrassment and everyone else's amusement. The incident taught me how little the country mouse knew of the town.

It was through my other engineering uncle, Uncle Alan, and his brother, my Uncle Jim, that I first came to know the Welsh Borderlands. Uncle Alan, when I first knew him, had already given up his career as a sea-going engineer and had settled down with his first wife, my Auntie Margaret, at the little village of Abdon-under-Clee (immortalised by A. E. Housman, of whom you will hear more), where she was the village schoolmistress. My Uncle Jim was the schoolmaster at the village of Pembridge (Figure 16) in Herefordshire, where he lived with my Auntie Elsie. Uncle Alan left Abdon when Auntie Margaret died and went to live literally on the Border at Hay-on-Wye. He kept a garage which, I believe, was the first building in Wales. I remember visiting Abdon only once, but Uncle Jim remained at Pembridge until I was in my twenties, and it was from there that we were principally able to explore the valleys of the Wye, the Teme and their tributaries in the Welsh hills. To go to 'another country' and actually see words written up in another language was less usual for small children than it is today, and to judge from the interest which our travellers' tales aroused among our friends back home in Stibbard we might have been to Samarkand or French Indo-China.

One other relation who comes into my story at this stage for a particular reason is my mother's youngest sister, Auntie Amy, who was also a schoolteacher, though she later abandoned the profession for librarianship. In the late twenties she lived in the schoolhouse at Marsworth, a village at the foot of the Chiltern Escarpment some thirty-five miles north-west of London. It was possible from the school playground to see the trains on the main line from Euston to the North-west, the same line which I had come to love so dearly at Bradley, a hundred and fifty miles further on. But the particular reason to which I referred was that it also lay at the junction of the Aylesbury Branch (or Aylesbury Arm) with the main line of the Grand Union Canal, still at that time actively carrying commercial traffic, some of it horse-drawn, close to the point where it was fed by four reservoirs. Here I was initiated into the *mystique* of another world, and when, thirty years later, I began to develop a serious academic interest in the inland waterways, it never ousted the emotional involvement which began in Marsworth. Whole families still lived on board the barges and narrowboats, and we talked to the children and peeped into their tiny homes. For them it must have been a hard life indeed, but for us on the tow-path it was the most romantic life imaginable.

As for the rest of my mother's family, her youngest brother, my Uncle Arthur, also lived in Newcastle with my Auntie Kitty. They had six children, Jimmy, Harold, Ruth, Dorothy, Alan and Eleanor, and I well remember some of them coming to Stibbard and camping with us at Weybourne on the Norfolk Coast, making the adventure of outdoor living something of a group activity.

My Uncle Jack lived for a time with us at Stibbard where he met and married my Auntie Mollie, yet another schoolteacher. They went to live in the schoolhouse at Ashley near Market Harborough in the East Midlands, and this became another frequent port of call, but at a rather later date when I was in my teens.

My mother's other sister, Auntie Jessie, had married another clergyman, Percy Opperman, and they lived in Dorset. I visited them once only during my childhood and I have a fairly hazy recollection of East Orchard, where they lived, but my cousins, John and Paul, used to come and stay with us at Stibbard. Their brother, David, was a good deal younger and my opportunity to get to know him well had to wait, paradoxically enough, until we were both middle-aged, by which time he had emigrated to Australia.

This account of my various relations is far too tenuous to give you any impression of them as personalities, but this was never the point of including this catalogue in my story. The message I wish to draw out of it is that, by the time I was nine-and-three-quarters, I had already at least superficially sampled a number of quite strongly contrasting landscape types, the basis of whose selection had nothing to do with the character of these landscapes but was to be found in an entirely independent set of circumstances, namely the series of separate decisions made at various times by my parents' brothers and sisters when they went to live where they did! Had their several *curricula vitae* been different, my schedule of early landscape experiences would not have followed the pattern it did. I didn't realise the significance of this until much later, but the recognition, when it came, brought home to me how powerfully the differences in our individual habits of environmental perception are moulded by the hand of chance.

By the time a boy is nine-and-three-quarters he will have accumulated an enormous heap of raw material from which to build his own world. That raw material will be drawn primarily from his own direct experience of those limited parts of the world which circumstances similar to those I have just outlined have arbitrarily brought within his field of observation. But this primary material will have been supplemented by other information and other pictorial images equally arbitrarily acquired at second, third or fourth hand, or even more indirectly. Furthermore his potential for building his own concepts of the world from this augmented store will have been further enhanced by many other variables, his powers of observation, his capacity for understanding what he has observed, his memory, his intelligence, his inventiveness and his imagination; but, just as it has been said of a computer that the quality of what comes out cannot surpass the quality of what goes in, so it is a fact that the developing picture of a child's model of the world is inescapably constrained by the nature of the original input. It would be remarkable indeed if so much variation in our individual experiences of the world were to result in the building of one single pattern of environmental perception or one uniform set of preferences common to us all.

Into the Unknown

Having concentrated so far on those images which derived from my own direct experience of places I had visited personally, I must add a little about those indirect experiences which had been concurrently worming their way into my picture-building through a number of agencies, chief among which were, not unnaturally, my parents.

My mother had travelled relatively little; indeed never in her whole life did she set foot outside the British Isles. Nevertheless it's arguable that her role in widening my horizons in these early years was more influential than that of my father. She was the principal story-teller, not only the reader of books at bedtime but also the recounter of travellers' tales brought back from Africa by my Uncles Ted and Percy and by Auntie Jessie, all of whom had been there as missionaries, and by my sea-going Uncle Alan who seemed to have been pretty well everywhere. They had all travelled more widely than any members of my father's family, but my father himself had not so long ago visited foreign fields in the King's uniform as an army chaplain, so my earliest introduction to the wider world had a heavy bias towards the Middle East, or 'Near East' as it was still called, as seen through the eyes of an army officer.

In these circumstances it's perhaps not surprising that my general picture of that part of the world where Europe, Asia and Africa meet was put together out of a plethora of *vignettes*, little anecdotes and episodes which depicted either the archaic, almost timeless way of life of the indigenous populations or the unnatural routine of an occupying military force. Not infrequently the point of a story would focus on the interaction between the two. The saga of the Eastern Mediterranean was thus gradually unfolded with the help of numerous photographs of varying quality which tended to fix the growing visual image into a series of stereotypes. On the one hand were the locals, Hellenic, with their olives and goatherds, or Arab, with their tents, camels and flowing robes, as the case might be; on the other the officers' messes, with their primitive army vehicles looking like military versions of bread-delivery vans and interminable groups of posed, nameless, uniformed men, most of them dead before I was born, who meant nothing to me except in the re-incarnations which my father's descriptions vividly contrived.

Some of these brother officers, together with a handful of old friends from school and from Oxford, became as real to me as if they had been regular visitors to our house. There was Onslow, for instance, whose portrait stood on the sideboard throughout the whole of my childhood and adolescent years. The idea that he had once been a civilian hardly crossed my mind, though my father's acquaintance with him came in fact from Oxford and not from the army. As far as I was concerned he had only one suit of clothes, the uniform which he was probably still wearing when he was killed. The fact that I never knew his Christian name, even though he was one of my father's closest friends, perhaps tells us something about the customs of those days. Young men at Edwardian Oxford really did refer not only to their college 'scouts' and their fathers' gardeners by their surnames alone but to their fellow undergraduates also.

Chief in the archives of my father's war memories was an album of photographs of a pantomime production entitled *Aladdin in Macedonia*. I came to know all the characters, their costumes, their make-up and the ranks of the actors who played them. The incongruity of a situation in which the Sultan might wear three stripes under his robe while the beggar's shirt covered three pips on the shoulder entirely escaped me, though I was impressed to learn that the princess was a bombardier. There can be few people in the world apart from me whose image of Salonika is built around a Widow Twanky wearing army boots and a moustache.

There was one curious thing about this world beyond the horizon which would immediately distinguish it from the comparable world of even my own children and certainly that of the children of today. It was all in sepia; not even black-and-white, but sepia. Variation of shade was achieved as it might be by adding more or less milk to a cup of coffee. Colour-printing was, of course, a feature of the more expensive books, supplementing for special pictures the more usual black-and-white, but even in books sepia pictures were not unusual. So my father as a young man had invariably taken his holidays in sepia resorts, his friends all wore sepia clothes to match their sepia complexions; Oxford was a sepia university and, above all, the England into which I was born had just emerged from a sepia war.

The war, in fact, was of such recent occurrence that the rumble of artillery still seemed to ruffle the tranquil summer evenings of North Norfolk in the early 'twenties; the smell of gunpowder still

hung about in the sheltered corners of the garden. The war memorial
at the further end of the Church Walk was the central cultural symbol
of the village, and I have very early recollections of the Stibbard Brass
Band leading the corporate respects of the parish on November the
Eleventh at eleven o'clock precisely as the whole activity of a nation
ground to a halt for the Two Minutes Silence, whatever day of the
week it happened to fall on.

Apart from my father most of the ex-service men of the village
had been in France. I came to know most of those who had come
back and I read the names of the others every day when I went to
fetch the milk. So there were tales of France as well to be added to
my stock of images - tales and pictures, in glorious sepia. If you
imagine France as a rich landscape of hills and valleys, woods and
cornfields, vineyards and villages, forget it. France was a shellscape
of mud and shattered tree-stumps stretching away to featureless
horizons.

What my mother lacked in first-hand experience of the wider
world she made up for in her devotion to those places which she had
come to value, and through her descriptive skills she communicated
her enthusiasm to me. These same skills she displayed in the telling
of stories and the reading aloud of children's books. Some of these
books influenced my developing taste enormously but it was not
until I was grown up that the thought seriously crossed my mind
that, just as my first contacts with actual places depended on where
my relations happened to live, so a toss-of-the-coin decision by an
aunt or uncle as to what book to give to a nephew for Christmas can
be responsible for injecting into his emerging picture of the world
new ideas and new images with incalculable capacities for further
growth.

I was fortunate in having access to many books, some of which
have left me with mental pictures which have never faded. I
remember particularly the books of Helen Bannerman, written at the
turn of the century some twenty years before I was born. I'm sure
that *Little Black Sambo* must be proscribed reading for today's small
children. The fact that they inspired in me a real affection for
children of another race would not protect them against the charge of
paternalism which admittedly oozes out of every page. Then there
were the various books of nursery rhymes, copiously illustrated by
artists who could have had no idea how deeply their images were to
mould the tastes of a future adult.

Our edition of the *Just So Stories* was illustrated by the author's
own black-and-white drawings with a strong influence of what I
would now recognise as *art nouveau*. Comparing them with the real
world as I knew it, I was more impressed by the divergences than the
similarities. This curvilinear style was fine for depicting a djinn

coming out of a bottle but not so good for portraying landscape; so I was left to forge my own image of the turbid Amazon and the great grey-green, greasy Limpopo River, both of which looked rather like the River Wensum with a few native huts on their banks.

At the other extreme were those books which had been illustrated in colour with plausible, realistic landscapes. The tales of Beatrix Potter were a case in point. Although her domestic scenes, with little animals behaving and conversing exactly like human beings, passed beyond the fanciful to the farcical, they were invariably set in authentic Lake District landscapes modelled on places I had actually seen. Cosy little paths threaded their way through meadows, oakwoods and hazel coppices punctuated by inviting little stiles where they crossed the dry-stone walls. A convincing sense of space was achieved by a combination of the rules of perspective and changes in tonal quality. The pastel shades and hazy distances induced an element of mystery without destroying the realism of the scene.

Kipling and Potter both illustrated their own stories, but there were other books in our little library with colour pictures by artists who specialised in illustrating the works of other authors. The most influential of these in the formation of my own taste was the late Harry G. Theaker. He had a remarkable capacity for restricting the content of his pictures to a level of simplicity which posed no problems of interpretation to a small child, (unlike the *art nouveau* of Kipling), yet introduced a powerful romantic dimension largely through his background landscapes. If, at nine-and-three-quarters, the subtler message of Cervantes was still beyond me, I had already fallen in love with the landscapes of Spain through Theaker's *Don Quixote.*

Not all of the stories I had come to know by this time were in illustrated editions. Our Hans Andersen, for instance, and our Grimm had no pictures at all, so we faced a challenge to build up our own visual images out of components drawn largely from our own experience. That is why, as I said, Hansel and Gretel lived in Swanton Great Wood; but as I became more familiar with illustrations of other continental folk tales the landscapes of the Brothers Grimm became increasingly teutonified, and I had, in fact, discovered the feel and the atmosphere of the German forests long before I ever set foot in one.

One work which, I can see with hindsight, was particularly influential arrived in the bookshops just in the nick of time. The first edition of *Winnie the Pooh* appeared a few weeks after my visit to the Isle of Man at the age of six. I must have been introduced to it immediately, because I find it difficult to imagine a time when there was no Pooh Bear, and I can distinctly remember my mother

producing a new book with new stories about the friends of the
forest when *The House at Pooh Corner* came out a couple of years later.
Here again the landscape images formed in my own mind were
powerfully influenced by the illustrations, in this case black-and-
white drawings by E. H. Shepard, based on the scenery of Ashdown
Forest on the borders of Sussex and Kent. Even so, I was recently
surprised to discover, on looking at his map of the forest once again,
how fickle is the memory. The Six Pine Trees, whatever their species,
were not the Scots Pines I had so vividly remembered. I think, if I
had to name my favourite tree, it would probably be the Scots Pine,
and I dare say this is a case of a first, accurate impression being over-
printed, as it were, by what later became a preferred symbol.

Not surprisingly in a rectory I encountered Bible stories at an
early age and here again, unless they were illustrated, I was left to
build my own pictures. So the Wilderness, the Sea of Galilee and the
Mount of Olives all contained elements drawn from the landscape of
North Norfolk. I have a clear recollection of my mother reading me
bible stories on the front lawn at Stibbard, and if you want to know
what sort of a tree it was that Zacchaeus climbed to see over the
heads of the crowd I can tell you it was a laburnum.

There was one book which played a particularly influential role
in the formation of my landscape taste and, at least in part, for a
particular reason. It was not exactly proscribed; worse than that, it
was half-proscribed, half-censored, half-withdrawn from circulation.
We were allowed to read it and yet we weren't. It was the subject of
one of my mother's few psychological blunders and it furnished my
first clearly recollected example of counter-productivity. The book
concerned was an illustrated version of *The King of the Golden River*,
and it was only in middle age that I discovered it was by John
Ruskin, a name which in childhood would have meant nothing to me
but which came to mean a great deal later.

The cause of my mother's anxiety was, I suppose,
understandable. Some of the illustrations by A. H. Baxter, were, to
say the least, potentially disturbing to a young child. One of them
depicted a sword fight between two hideously ugly brothers, Hans
and Schwartz, and I dare say the idea that such a relationship could
even be contemplated might be thought to present emotional
difficulties to a little brother and sister! Admittedly the faces of the
contestants were pretty frightening and the reason given for the
prohibition was that these illustrations would make us dream, so we
were allowed access to the book only occasionally and never just
before bedtime. A more certain recipe for the development of a
phobic fascination is hard to imagine!

You may wonder why, if the book was judged to be so
unsuitable, we were allowed access to it at all. The answer is that it

FIGURE 17. FORBIDDEN FRUIT. An illustration by A. H. Baxter from *The King of the Golden River* by John Ruskin. The caption reads: 'My son', said the old man, 'I am faint with thirst'. The original is in colour, but the dramatic scenery with its 'sublime' precipices, waterfalls, etc., can be clearly seen in this black-and-white version which is reproduced by courtesy of *Geographia*, London.

was a present from Uncle Edge, my father's eldest brother, who, having been christened Charles Egerton, anticipated the practice of most present-day Australians and some others by going through life, all ninety-five years of it, under the label of the first syllable of a Christian name. I later came to realise that *The King of the Golden River* was just one of a number of issues which had collectively moulded my mother's attitude to Uncle Edge. It was not an attitude of hostility or aversion; on the contrary, I believe she felt quite an affection for him in her own way, but it was an affection mixed with a large pinch of exasperation. She regarded him, in short, as mildly irresponsible in a way which could be both amusing and infuriating.

It was this same Uncle Edge who had so generously defended us when we laughed at his bee sting. It was he who used to take my part when I wanted to go and see the trains instead of dangling my little legs while the grown-ups finished their lunch. Not every small boy had a solicitor to plead his cause at table! It was he, too, who had first told us the story of the Welsh Knight, alleged to have been slain near Newton-le-Willows while on some ungentlemanly errand involving one of the young ladies of Lancashire, and who had persuaded us that it was not only permissible but our patriotic duty to express our contempt by spitting at the stone which marked the place where retribution had at last caught up with him. You can imagine that the inside of a car, or worse still the side-car of a motor-cycle combination, was not the best place for such a gesture of disdain, and I suppose this didn't help to foster the image of Uncle Edge as a good influence. It was he, too, who later constructed the little railway, and my mother must have had many an occasion to be thankful that it remained a monument merely to my uncle's inventive eccentricity and not to a couple of dead kids.

From all this you will see that, if anybody had to go and give us a book with horrible pictures, it had to be Uncle Edge. As my mother might have said if the phrase had been invented, 'He would, wouldn't he?' So she was left with this predicament; either she would have to organise controlled exposures to the risk of nightmares or think up an explanation for our ignorance of the book's existence!

This devious path brings me at last to the point of the story, which is that, as well as the horrible pictures, the book also contained some dramatically romantic landscapes, one of which is reproduced as Figure 17. When I pestered my mother to let me have one more look it was really this which had captivated me. The little rivulet came like a thread of gold out of the sky, as it were, and cascaded in a series of plunging arcs down, down, down into the mysterious forests of the Styrian Alps. What a landscape! What a book!

Interesting as they were, photographic illustrations of the real
world were by contrast prosaic and mundane (particularly if they
were in sepia!). They lacked magic. Nevertheless some of them had
a magnetism of a different kind. I remember a series of booklets
called *An Outline of the World Today*, which was a kind of introduction
to the geography of far-away places. They contained some quite
memorable pictures. I remember particularly the Canadian Houses
of Parliament in Ottawa and the famous bell tower in Bruges, but
nobody ever thought to lock *them* up at bedtime. I suspect they must
have been used by my mother in a thoughtfully devised programme
of instruction about the world beyond the horizon.

It was within this programme that she introduced me very early
to the atlas. I can see her now sitting under the pink chestnut tree
with the atlas on her lap, fitting various pictorial information into the
various countries, fitting in also those verbal descriptions of places
visited by my father in the Eastern Mediterranean, by my Uncle
Alan, who, as a sea-going engineer, had sailed up the Amazon,
('Now, shall we find the Amazon?'), and by my Uncles Ted and Percy
and my Aunt Jessie who between them had worked in Central
Africa.

The missionary societies provided for us an important window
on the world, not least through the exhibitions which were staged as
educational-plus-fund-raising exercises in the various market towns
of North Norfolk. My father often used to get involved in the
organisation of these and Helen and I were taken to a good many.
The effect was an exciting and enjoyable bombardment of
information which my intellectual processes were not mature enough
to cope with in any systematic way, so there resulted a glorious
confusion of images which, decades later, I attempted to express in
verse.

THE MISSIONARY EXHIBITION

A little boy's kaleidoscopic mind
Creeps through the jungle of a Norfolk town
Breaking the great *apartheid* of mankind
Victoriously down.

Here grassy skirts tempt decorated spears,
And here the tweed-encumbered Mrs Giles
Re-lives her forty bible-clouded years
Among the crocodiles.

Bright, varnished paddles rest on varnished boats,
While varnished boxes, scattered here and there,

Like hungry nestlings stretch their varnished throats
For numismatic fare.

African chiefs greet Chinese mandarins
In cardboard characters, and over all
The cold, didactic smell of drawing pins
Invigorates the hall.

Above the rows of colour-printed huts
Under a palm tree's azimuthal crown
Clutches of pinhole-punctured coconuts
Glance perilously down.

Somewhere behind the road to Mandalay
Gas-rings emitting periodic pops
Proclaim some high domestic holiday
Back-stage among the slops.

In tea-stained saucers stacked on tinny trays
Witch-doctors stub evangelistic fags
And pile up coins on moth-corrupted baize
For stuffing into bags.

And last, the brightest memory of the lot!
At Coltishall in Nineteen-twenty five
They boiled a missionary in a pot
There in the hall - alive!

(Appleton, 1978:31).

From all this you will see that the wide world beyond the horizon was by no means a closed book, but it had to be reached through different channels from those which led to an understanding of those parts of the world which I had been able to see for myself. Pictures of various kinds, together with the printed and spoken word, provided the avenues through which this knowledge could be attained, but the penetration of those avenues required a versatile vehicle, and that vehicle was the imagination. Only through the imagination could this huge heap of raw material be converted into something manageable.

By the age of nine-and-three-quarters, then, I had circumnavigated the globe, steering my little ship from the captain's quarters in the nursery cupboard. I had crossed the Pacific many times under the kitchen table on which had been placed books and other heavy objects to hold in position the rugs, towels and table-cloths which draped to the floor and closed in the cosiest of cabins, well-stocked with sweets, biscuits, lemonade and other necessities indispensable for an ocean-going mariner. I had crossed the Antarctic Ice-cap in a sophisticated sledge of my own design. The inside, luxuriously appointed with cushions, blankets and pillows, was furnished from my mother's stock of household effects. The vast expanses of snow and open sky outside were furnished, like the engine, from my own imagination. The thing I most clearly remember about the adventure is the sensation of being enclosed in a safe, cosy place and looking out at a potentially hostile world in the confident expectation that I could overcome every impediment which Nature might throw in my path and eventually attain my objective of placing my tiny Union Jack on the South Pole.

So the uttermost parts of the earth became accessible; the unattainable became attained. By the nineteen twenties space fiction had already carried numerous heroes to the moon, and the fact that nobody, as we all knew, would ever be able actually to walk on it didn't preclude the enjoyment of the impossible. So we went to the moon too, a delightful moon with an atmosphere just like our own and even human inhabitants, not only friendly but even English-speaking. At nine-and-three-quarters one didn't have to stop at the moon either!

But what's so special about nine-and-three-quarters? For you maybe nothing; for me it was a watershed. In many ways I'm sure I was the same little boy at ten as I had been at nine-and-a-half, just as in many ways the England of 1067 was the same as the England of 1065; yet we all know that 1066 was literally epoch-making, and what 1066 was to English history nine-and-three-quarters was to me. In September 1929 I was sent away to school.

3 THE INSTITUTIONAL LANDSCAPE

Shades of the prison-house begin to close
Upon the growing boy.'

Sicut Aquilae

The Glebe House Preparatory School consisted of a couple of rather gaunt three-storey buildings of ginger carstone on the edge of Hunstanton, a little seaside resort tucked away in the north-west corner of Norfolk some twenty miles or so from my home at Stibbard. It was a small school run by a headmaster who had a tolerable grasp of the classics and a very fair smattering of general knowledge but not, as I now think, a very deep understanding of children, which admittedly was not a qualification universally expected of schoolmasters in those days. When he had a good idea it was not always matched by the sort of imagination necessary to make the most of it. I'm all for introducing children to good music at an early age, but can you imagine rows of little boys, none of them over thirteen, sitting on a Sunday evening listening to complete acts of Wagner operas?

Anyhow, one of the things he taught us, which has stuck in my memory, is the meaning of the phrase *de mortuis nil nisi bonum*, 'speak nothing but good of the dead', and in retrospect I can't help wondering whether this was a self-protective measure of unusual far-sightedness to immobilise the word-processors of his little charges half a century later. In any case I have warned you that this is not really a book about people, except in so far as they affected my habits of perceiving the world around me, and with regard to his personal shortcomings, about which I admit to a probably unfair prejudice, I will let him off the hook and concentrate on the main theme.

School, they say, is a way of life, and it's questionable how far one can pick on selected aspects of the total experience and attribute to them specific influences on one's own subsequent patterns of behaviour. But there were some aspects which I feel fairly confident in identifying as being at least in part responsible for my later tastes and preferences.

I had, of course, become accustomed to the discipline of doing formal lessons at Ashworths, (and incidentally the trauma of leaving home for the first time was made more tolerable by the fact that John Ashworth, though a year younger than I, also went to the Glebe House at the same time), but lessons with Miss Marler had only lasted for a few hours. For the rest of the day I had been pleasurably free to do my own thing within pretty wide constraints. Exploration had always loomed large in my system of preferences, and now I found myself severely limited in both time and space. Every moment of the day, weekdays and Sundays, was planned by somebody else. Many of these activities were quite enjoyable, but they were not of my own choosing; I was becoming a puppet, manipulated by people who seemed to have little comprehension of my own wants and needs. Procrastination, I was later told, is the thief of time, but at the Glebe House it seemed to be my teachers who were stealing time, my time, and depriving me of the use of what I thought properly belonged to me.

Spatially, too, I was no less confined. All the activities which were considered legitimate - or almost all - could be accomplished within the school's buildings or its grounds. I say 'almost all'. Twice a day on Sundays we were permitted, indeed compelled, to venture out into the real world. In the morning we went to church at Old Hunstanton, about a mile away, where the Headmaster read the lessons (one of the credentials which I believe persuaded my parents to send me there in the first place). He invariably arrived before us in his car while we were conducted in crocodile formation by one of his staff. In the afternoon we set off in the same formation and at first in the same direction. In about half a mile, however, instead of turning right towards the church, we continued straight ahead to the motley collection of houses and bathing huts near the Golf Club and made our way down on to the beach.

It was here that we achieved our nearest approach to personal liberty for the space of an hour or so. We could even play, as long as we didn't become too boisterous or in other ways let down the tone of the school. We crawled under the beach huts which were built up on posts a couple of feet or so in height, rather in the style which decades later I was to discover is so widely practised for excellent environmental reasons in Queensland, where they have snakes and termites and poorly circulating humid air. We ran races on the sand; we fought guerrilla campaigns in the dunes, seeking such strategic cover as might be obtained from the spiny clumps of marram grass.

In the other direction we were sometimes allowed under escort to walk below the cliffs where the chalk came in two colours, white at the top and bright pink at the bottom, both making a striking contrast with the surmounting carpet of green grass. Great boulders of this pink chalk lay nine-tenths buried in the sand of the beach. Many years were to elapse before I was to see more striking colour contrasts in nature.

The hour of liberty soon came to an end and we formed up to be shepherded back in our school caps to the confinement of the two gaunt buildings and the resumption of the normal routine of approved and supervised activities. The route was always the same. In four years it never varied once, so we had to find novelty in the changes of the seasons as the trees acquired and discarded their leaves and the fields nurtured their crops and gave them up to harvest, or in the changes of light which added a further variety to this otherwise limited landscape menu. It may be that the very deprivation which I so resented laid the foundation of a heightened sense of values to be found in my surroundings simply as objects of visual perception, a heightened pleasure in the very act of perceiving.

Of all the changes I experienced in going from home to school the one which took most getting used to was the lack of privacy. At home it was always possible to retreat from the Outer into the Inner Bailey, from the Inner Bailey into the Keep and finally into the inner recesses of my own bedroom where even my own mother would not dream of entering without knocking on the door. There I could be assured of seclusion from the intrusive eye of all humanity. It was totally my own world. Whether I shared it with anyone else was my own decision. To have to share one's own bedroom with five or six other boys not even of one's own choosing was like being forced to strip naked in a public place. The bed itself was the only really private retreat. I used to tuck the blankets in so tight that they drew the sides of the mattress together like the hull of a boat. It was like going to bed in a kind of tube, and there I cocooned myself in a little world of my own to await whatever violations of my privacy the morning might bring.

Perhaps the ultimate such violation was being made to eat all the food that was put in front of me. Admittedly I was a very faddy little boy when I arrived there, and I suppose it might have been argued that an enforced familiarity with new tastes and new textures would expand the list of foods in which I would eventually take pleasure. In fact it had the opposite effect. I became neurotically anxious about putting anything unfamiliar into my mouth. I wasn't, of course, force fed, but I wasn't allowed to go and join the others in their precious moments of spare time until I had finished, and this would be a protracted business. There were some foods which I simply couldn't swallow without retching or even vomiting, except in minuscule quantities washed down with copious draughts of water, a technique which, though eventually successful, brought problems of another kind in the ensuing lesson or on the football field.

I shall refer later to the influence of the curriculum on my developing tastes in landscape and on my attempts to explain theoretically the aesthetics of environmental perception. Here it will suffice to single out just three subjects which steered my thinking habits in important directions. The first two were foreign languages, which at this stage meant Latin and French. These we learnt by methods which had been well tried during the nineteenth century. We started straight away with declensions and conjugations, that is to say studying language as a science. The idea of using language as a vehicle for the acquisition of information and the expression of ideas had to wait quite a long time. When it eventually came it made the study of French a good deal more interesting. We collectively read *Ma Première Visite à Paris* in which we encountered named individuals who faced credible situations in plausible ways. We even took a boat down the Seine and climbed the Eiffel Tower, and there began to emerge a rudimentary picture of how the street plan of the French capital fitted together.

But in the Latin Primer the facts and the ideas were so far removed from my own experience of the real world that I found little compensation in the content of the words. The fact that Caesar, by forced marches, reached the camp with two cohorts at first light made no visual impression that related to the world which I had been discovering under the hedges and in the ditches, in Ashworth's hayloft or in Stibbard churchyard or on the road to Bradley. Nobody ever thought to tell me what a cohort was, and I probably wasn't interested enough to ask. As for a forced march, all I knew was that *magnum* meant 'great' and *iter* meant 'a journey', but that, if you found them together as *magnum iter,* you had to translate it as a 'forced march'. There was no mental picture of reality, no straggling lines of exhausted soldiers, no sweat-sodden shirts, no chafing of coarse military fabrics on inflamed skin, no oaths, no obscenities, no raging unquenchable thirst, no sense of endlessness under a burning Mediterranean sun.

The only gesture made towards the apathetic young reader was to illustrate the text with simple line drawings of figures from the ancient world. More often than not the feature which caught the imagination and therefore had the best chance of prompting a question, articulated or not, had little or nothing to do with the point of illustration. Why did the vase have a piece missing? Why did the lady have no arms? Why did the old man with a beard have eyeballs but no eyes? Small wonder that most of our books had copious supplementary drawings in the margins, crudely executed in pencil, or, more injudiciously, in pen-and-ink, and illustrating whatever extraneous thoughts happened to be passing through our minds in the middle of our lessons.

The third subject which was later to emerge as important in my life was geography. By modern standards the method of teaching must have violated every rule in the book; it would have sent an educational theorist into convulsions if indeed he or she could have been induced to believe what we were required to do! But we aren't judging by modern standards, so that's neither here nor there. We were each given a book containing a number of black-and-white outline maps of the several countries and continents of the world on different scales together with associated lists, indeed tables of place names - rivers, mountains, deserts, lakes and towns, not to mention the proverbial capes and bays, all of which we had to mark on the maps. We were then supposed to learn their locations by heart so that we could repeat the exercise with the highest degree of accuracy in the examination at the end of term.

Old-fashioned or not, by contrast with Caesar's forced marches, Canning's foreign policy and long division I found all this simply captivating. The basic techniques of map-making, to which my mother had introduced me, gave me a head start over my fellow-pupils, and no doubt a part of the satisfaction which I found in geography came from my comparative successes. I once scored ninety-nine percent in a geography exam which had to be compared not only with the fifties and sixties which were my best scores in other subjects but also with the thirties and forties which my rivals were scoring on the same paper. They, needless to say, found the whole exercise perfectly ghastly, perfectly soul-destroying and perfectly pointless, whereas I could at least understand what it was getting at. The representation of places which bore the same relationship to each other on a piece of paper as did their prototypes on the surface of the earth was at least an intelligible concept which seized my imagination. I might not know who, if anybody, lived in the Gobi Desert or among the peaks of the Andes, but I certainly knew where they were and could reproduce the evidence with black lines for mountain chains punctuated where appropriate with little open circles for their major summits, and somehow it all made sense.

At the beginning of my last year when I was twelve I had to give up geography so that I could devote myself to the more serious business of learning Greek. It was like starting Latin grammar all over again only with the additional trauma of funny letters, and if I have painted my recollections of geography in more glowing colours than the facts strictly merit you may think of me as you might think of some love-lorn swain deprived of the maiden of his dreams and unable to think of her except as a symbol of perfection. Geography, I suspect, shared the fate of *The King of the Golden River;* it had been proffered, it had been tasted and it had been withdrawn, but it had not been forgotten.

Perhaps one of the reasons why geography so endeared itself to me was that it was so manifestly concerned with places far away, and therefore outside the boundaries of the school, in a way that was not quite true of other subjects. Chemistry, for instance, which paradoxically was taught in quite an imaginative and interesting way by an octogenarian, was based on experiments which, like the theorems of Euclid or fourth-conjugation verbs, were not place-specific, and could be thought of as something which properly belonged to the classroom. But the Amazon didn't really mean anything except in the context of South America, and if the book didn't tell me much about it except where it was, hadn't my Uncle Alan sailed up it, and hadn't my mother told me stories of its forests and of the people who lived on its banks and of the nasties that swam in its waters? I had even learnt from Kipling that it was 'turbid', whatever that might mean.

I think it was this concept of a vast freely intercommunicating world outside the school gates which high-lighted the all-too-real dichotomy between 'inside' and 'outside'. The symbolic image of the prison was not so far-fetched. My territory was delimited by the boundaries of the school grounds. On Sundays we were taken out under supervision as my sister and I used to take our guinea-pigs out of their cages for a brief period of liberty to graze on the lawn under our watchful eyes. But they knew well enough that their proper place was inside. If they looked out on the stick-heap and the snowberry bushes and the little cobweb shed they must have seen these things as unattainable objects in an inaccessible landscape beyond the wire netting, a landscape in which they had no part to play themselves, and that is exactly how I saw that little corner of North-west Norfolk beyond the peripheral fence of the Glebe House.

There were, in fact, two ways in which I was able to reach this magnetic world outside the prison wall in addition to those periods of physical release on Sundays. The first was through direct visual perception, the view from the window; the second was through the imagination. In the first place there were a few opportunities within strict limits to perceive this wider world literally and directly. The buildings were tall and at least some of the dormitories were on the top floor. From their windows one could command a wide view over the gently rolling countryside and, to the north and west, the sea. The County of Norfolk projects in a large bulge on the eastern side of that shallow arm of the sea known as The Wash, and for this reason at Hunstanton, although it is on the east coast of England, the sun may be seen setting over the sea.

One year I found myself in a dormitory with a window which looked out in that direction. On the wall beside it there hung a picture of a horse and rider entitled 'Bellerophon', or was it 'Bucephalus'? Whether this was the name of the rider setting out to slay the Chimera, (it could hardly be Pegasus because it had no wings), or of the favourite mount of Alexander the Great is unimportant. As far as I was concerned the horseman was me, heading out into the perpetual sunset, beyond the trees near the school gate, beyond the ginger sandstone façades of the town, even beyond the glistening sea into the low streaks of coloured cloud which seemed to contain the spirit of the dying light. It was, I think, the most powerfully romantic vision I had experienced up to that time.

For all practical purposes the real world was what I encountered every day in the classroom and the dining hall and on the football field; the world of the sunset was a dream world, beautiful but illusory. But suppose it should be the other way round! Surely what I could see through the window was equally a part of the real world! With a little more effort, a little more imagination, a little more *belief* it should be possible to dethrone the school and set up in its place the world of the perpetual sunset. Why, if it were not true, should I find myself kneeling there on my bed for quite literally hours staring out through the long, light summer evenings and feeling so intensely that in some mysterious way that was where a part of me belonged? Although it was far beyond my capability to rationalise it, I now understand that what I had come to believe in so passionately was some system of absolute beauty inherent in what I could see out there in the evening sky, from which I was, for the time being at least, separated physically though not emotionally.

This was to be for years the basis of my aesthetic philosophy, and only in middle age did I find an alternative explanatory framework which was able not so much to replace it as to contain it. The reason for my referring to it now is that this little story of Bellerophon (or Bucephalus) and the sunset establishes that by the age of eleven at the latest (after that I slept in another building), I had discovered that this power of attraction towards some quality apparently inherent in nature had become as real to me as the environment itself, perhaps even more real. Just as the personality of a loved one seems to lie behind the face and to find expression through it, so there was something which lay behind the sunset and behind the poplars stretching up into the blue-and-white sky by the meadows on the way to Ashworth's farm and behind a thousand other experiences of the world which seemed to contain some meaning or value which I couldn't describe but which found expression through all these material objects. It was to be some years before I discovered that Plato had been playing around with these ideas over a couple of thousand years ago, a fact which, had I known it in 1931, might have put me off the sunset for life!

Once I had accepted that there were forces at work which I could not identify but to which I responded emotionally, it became possible to imagine some sort of relationship between the feelings aroused by nature and the feelings aroused by other things, like music and even poetry. I shall return to this subject later, but at this stage I should like to record one incident which in retrospect seems to have been a milestone.

There comes a period in most schools after examinations and before the end of term when formal lessons are abandoned and some other programme of approved activities is laid on instead, probably to keep the children semi-usefully employed while the teachers get on with marking the papers. On one such occasion an assistant master sought to keep us occupied by reading to us Fitzgerald's translation of *The Rubáiyát of Omar Khayyám*. We subsequently heard a rumour that this was not approved by the Headmaster, a rumour which I half hoped was false because I liked this master and didn't want him to get into trouble, and half hoped was true, because that would have been the ultimate accolade and confirmation that he was the good fellow I thought he was. At all events it was a revolutionary experience for me, and for at least three reasons.

First, it brought home to me, perhaps for the first time, that metrical rhyming verse is a kind of music, making a direct impact on the senses through patterns of sound. I'm sure I wouldn't have expressed it like that; let's say I just wanted the sensation to go on, an unusual response to most classroom activities, even the study of geography!

Secondly, it confronted me with a confusingly ambivalent challenge. I was still at this stage committed to a fairly simplistic view of a theology in which a benign God manipulated the day-to-day affairs of a subservient humanity, and here was somebody apparently calling in question the very essence of the religion in which I had been brought up at home and which was still being actively instilled into me at school. What I enjoyed was not being disillusioned, because I don't believe I was, but finding myself in sympathy with a poet whose message I was able to reject rationally while allowing myself to be swept off my feet emotionally.

The third reason why I think this was an important episode in my life was that it started me making a more serious attempt than any I had made previously to write rhyming metrical verse myself. It wasn't any good, but it did rhyme and it did scan and it taught me that verse isn't just something to be taken in, like the vision of the sunset, but that it can also be given out as an expression of oneself. If Dudley Biggs is still alive and should happen to read these words I hope he will allow me to thank him. He might even tell me whether he did get into trouble!

I completed my four-year sentence at the Glebe House in July, 1933. It had set out to inculcate a reverence for all the Cs, for my Country, for the Crown, for the Constitution, for the Classics, for the Conservative Party and for the Church of England. Success was patchy. The school motto, *Sicut aquilae*, was taken from the fortieth chapter of the Book of Isaiah and the thirty-first verse. 'But they that wait upon the Lord shall renew their strength; they shall mount up with wings **as eagles**; they shall run and not be weary; they shall walk and not faint.' It was intended, no doubt, to be a sort of stoic exhortation to probity, fortitude and effort, but for me the symbolic meaning of the eagle was rather different. It was a strong bird with powerful wings on which it could soar into the sky over the fence and high above the trees, and fly away, far, far away, into the sunset.

Floreat Salopia!

I'm tempted to suggest that Shrewsbury School was to the Glebe House what an open prison is to a top-security gaol; but that would be unfair, so I won't. As at the Glebe House, and presumably most schools, the daily round continued to be planned according to a rigid timetable, but there were much longer periods of free time and greater opportunities to do one's own thing. Most important for our purposes, however, were the very different rules which governed our freedom of movement. For most of the day we were generally confined to the school grounds (The Site) more by the location of the activities in which we were engaged than by any direct prohibition. In any case 'The Site' itself was quite large in area and comprised widely contrasting visual components. We were accommodated in ten or a dozen houses, each containing on average about fifty boys (invariably referred to as 'men'), though the one I lived in was twice the size of the others and therefore divided for games, etc., into two. Some of these houses were on the other side of the public road as also were the houses of many of the masters which we sometimes had legitimate occasion to visit. The sense of total confinement, therefore, which had so overwhelmed me at the Glebe House, was replaced, to my great relief, by a sense of comparative liberty.

Unless we were given special permission to visit it for some particular purpose, like buying a present for an aunt or visiting the dentist, the town itself was a no-go area. The usual explanation for this was that the school had to protect itself against epidemics caused by the germs we would pick up in cinemas and such places, but it was widely believed that it had more to do with certain naughty ladies who were reputed to live somewhere beyond the English Bridge. Be that as it may, 'down town' was out of bounds to any but the upper *échelons* of the hierarchical order.

The town itself lies in a loop of the River Severn, and since 1882, when it moved out to a new site, the school, invariably referred to in the plural as The Schools, had occupied a part of the plateau of Kingsland beyond the river to the south-west of the town. Further out still lay the recently constructed Shrewsbury By-pass which carried the A5 Trunk Road round the town on its way to North Wales. Beyond this and its extension eastwards to Much Wenlock in the south-east we were free to venture as far as we liked through approximately a hundred and eighty degrees of arc provided we returned in time for the next scheduled engagement. Once across the By-pass one quickly reached open country and every Sunday afternoon I would set out to explore its lanes, its footpaths, its fields and its streambeds, more often than not in the company of Robert Hudson, later to become one of the better-known commentators of the B.B.C., and its Director of Outside Broadcasting (Hudson, R. 1993). We shared a taste for the rural landscape.

Apart from having to wear ridiculous clothes like Eton collars and straw boaters with house-colour hat-bands, it was almost like being back in the by-ways and hedgerows of Stibbard. Our favourite venue was a place called Pulley Common. It has, alas, been ploughed up long ago, victim, I suppose, of the wartime campaign to grow more food, but in those days it spread like a green blanket down to the murky waters of the little Rea Brook. It was intermittently covered with hazels and hawthorns, brambles and blackthorns, muddy places and scruffy little patches of open grass hiding among clusters of self-sown vegetation. The very idea of being *allowed* to go to such a spot seemed inconceivable after four years of confinement, so not surprisingly we made the most of it.

Our destination wasn't always Pulley Common. Sometimes we would go over Sharpstone Hill to the woods at Bomere. This was an extraordinary place. It contained two small glacial lakes, Bomere Pool and Shomere Pool. They were encircled by steep wooded banks which so sheltered them that one could often see dark reflections in their still waters. They exuded a kind of gothic gloom and could easily have come straight out of a novel by Mrs Radcliffe. This was witch's country, alien and mysterious, no more than three miles from the middle of Shrewsbury, yet to all appearances belonging to a different world.

With increasing seniority I was allowed a bicycle and this immediately increased the scope of my travels. I shall tell you more

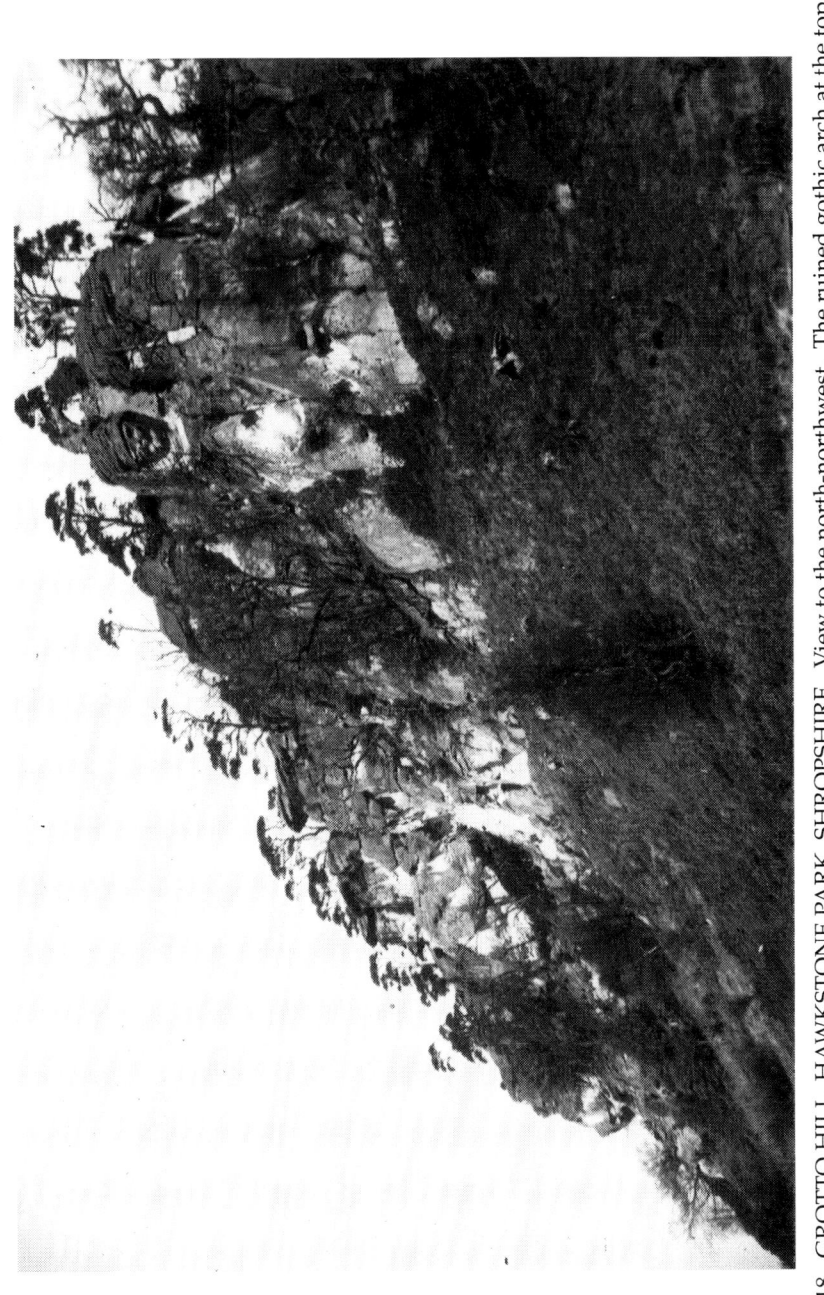

FIGURE 18. GROTTO HILL, HAWKSTONE PARK, SHROPSHIRE. View to the north-northwest. The ruined gothic arch at the top of the precipice of Keuper Sandstone marks the position of the grotto. This is arguably the most dramatic expression of the late eighteenth-century 'Picturesque' style of landscape design to be found in Britain. Photo by the author, *circa* 1970.

later of the compulsive interest in railways which played an important part in my life, and, since this was something of an acquired taste, not shared by any of my immediate circle of friends, I found myself increasingly exploring the railways of Shropshire alone. I had also begun to form strong preferences for certain kinds of environment and particularly for the hills. Shrewsbury itself lies in a broad shallow basin through which the River Severn pursues a sinuous course and from which it eventually escapes by way of the Severn Gorge; but this lay east of the Much Wenlock road and was therefore out of bounds. Much later, when my job required me to conduct field courses for geography students, this Mecca of the Industrial Revolution was to become a much frequented place of pilgrimage.

The southern and western sides of the Shropshire Basin were enclosed by various hills to which also I shall refer later. Their smaller outliers were not more than eight or nine miles from The Schools and I often cycled as far as the nearest of them at Pontesford. It's a central theme of this book that our individual opportunities to experience particular kinds of landscape are conditioned by fortuitous circumstances, and, had it not been for the attitude of my headmaster, those higher hills, which formed such a spectacular skyline to the south and west, might have remained just that, a skyline.

H. H. Hardy, however, was a passionate devotee of the hills and a very competent alpine climber. On occasions I believe he took specially selected boys climbing on Snowdon, but that was too strong a medicine for me and I was more interested in the gentler rambles which he organised. Once a term a whole day would be set aside for what he called 'country expeditions'. All school work was cancelled and we were expected to sign on for one of the options on offer. Certainly these were conducted, supervised affairs, but one could readily put up with a measure of organisation and control if the reward were to savour the delights of being able to become acquainted with new, more distant and more exciting places. There is a tremendous richness of landscape accessible from Shrewsbury, and I would return from these country expeditions with indelible memories of an enchanting countryside and a kind of mental album of views of the distant Welsh mountains as seen from numerous vantage-points.

Of all the places I visited in the vicinity of Shrewsbury perhaps the most memorable was the dramatic landscape of Hawkstone Park. I had at this time no knowledge whatever of the history of English landscape gardening and many years were to elapse before I realised the significance of this place in the evolutionary history of landscape style and fashion. It was laid out during the last two decades of the eighteenth century at a place where bands of sandstone, outcropping among the surrounding shales, provided spectacular red cliffs and it embodied uniquely the spirit of the Picturesque. Its most spectacular feature was a grotto gouged out of the sandstone near the top of a pine-forested cliff (Figure 18). In the nineteen thirties it was still in pretty good condition. The cave was lined throughout with shells and brightly coloured minerals placed there by the ladies of the household as an extraordinary expression of a specialised art. It's now in a state of lugubrious dereliction, in a way even more evocative of the spirit of the Picturesque, mantled as it is by a cover of untamed vegetation, though, as I write, plans are under discussion for its renovation. In any case there remains with me the memory of the place more or less as its creators intended it to be, and that first visit over fifty years ago was by any criterion one of the important moments in my experience of landscape.

There was one other institution under whose auspices we visited the surrounding countryside, the Officers' Training Corps, in which I enlisted as everyone was expected to. Most of our activities consisted of parades on the school site with occasional short route marches along the surrounding roads and even more occasional visits to the shooting range at Sharpstone Hill; but approximately once a term we had a field day and during the summer holidays there was always a week's camp. Owing to illness or some other acceptable excuse I attended only once, at Tidworth, but this was enough to introduce me to the hitherto unfamiliar landscapes of Salisbury Plain.

The field days held during the term usually started with a route march to Shrewsbury Station which was on the opposite side of the town. A special train would then take us to some remote railway station, of which there were then many more than there are now, and there we would detrain and take up our battle positions. The officers were all masters but boys could reach the higher non-commissioned ranks if they were sufficiently keen. Some of them intended to make the army their career and they took it all very seriously indeed. A boy in my house finished up as a general, but most of us lacked the same degree of commitment. The serious business still lay a few years ahead, but we didn't know that at the time, and we entered into the spirit of the thing as we saw it, a chance to get away from school for a day to play an enjoyable game.

Games, however, can be very realistic, and as we crawled forward on our tummies trying to remember what we had learnt from the manual of infantry training about five-barred gates and bushy-topped trees it wasn't difficult to imagine that the fire being directed from the opposite hillside was coming at us, not from blank ammunition, but for real. For me this was a new concept of landscape. I had of course played at soldiers on the beach at Old Hunstanton and before that at Stibbard. Had I not manned the ramparts of the Inner Bailey and even directed withering fire in the shape of a syrup tin lid at the wood-shed door from the little fort in the shrubbery? But now rather less was left to the imagination. We learnt what was meant by a commanding position and why we had to be particularly aware of dead ground, that is to say ground which could not be seen from a particular vantage-point, and it was not hard to understand that, in certain circumstances, environmental perception and biological survival might amount to much the same thing.

There was just one other kind of exploratory experience which I will mention before passing on to other matters. The school had a very flourishing Boat Club. It competed every year at Henley Regatta and proved a regular nursery for oarsmen in the Oxford and Cambridge crews. My father had been Captain of Boats and might well have obtained his Rowing Blue had it not been for that accident to his wrists. But for some of its members, and I was one of them, the attractions of joining the Boat Club were social and pastoral rather than athletic and competitive. Membership carried with it the privilege of going 'up-river'. Two or three of us would take a 'clinker pair' or maybe even a couple of canoes, a pound or two of sausages and a frying-pan and head off upstream. The river here ran through territory which, according to the school rules, was otherwise out of bounds. There was a Jerome K. Jerome dimension to the whole exercise, and no doubt this was part of the attraction, but there was more to it than that. There is something quite unique about penetrating an otherwise inaccessible piece of country by water. The river was nature's primeval highway, and for mile after mile above Shrewsbury the Severn sweeps in wide meanders which are never reached, much less crossed, by a public road. For a part of England which has been settled and farmed for over a thousand years there's an unusual isolation about its secluded banks on which, at a point of our own choosing, we would make a fire from whatever materials nature provided. An up-river picnic is one of those experiences of more than half a century ago that I would dearly love to revive.

You may well be wondering, while all this exploration was going on by foot, by bicycle and by boat, what was happening in the schoolroom. The short answer is 'the mixture as before', only rather stronger. French, Latin, Greek, History, English Language and Literature, Maths, Chemistry, Physics, Biology and, of course Religious Knowledge, which, however, was known as Divinity. And Geography? Alas, no! Geography was taught in the Third Form, which was the lowest in the school, and I believe in the Lower Fourth, but I missed these out. Really bright boys went straight into the Fifth. I started three rungs on the ladder below that, but even that modest elevation was exalted enough to save me from what was generally regarded as a just punishment for sloth. Geography was certainly a School Certificate subject and was therefore a sort of lifeline for those who couldn't be expected to pass in anything else. But having been placed in Middle IVA I had been pronounced capable of better things, so geography was not for me. As for that ninety-nine percent, I had better forget it!

For five years, therefore, I slogged my way through the Classics. My cousin, Alfred Mather, who was in another house, and I sat in the back row. Together we studied the wars of Scipio against Carthage, ironically unaware that Alfred's own life was to end all too soon in another North African Campaign. It was neither a very easy nor a very rewarding experience. I never really mastered the basis of Greek grammar, and I suspect that one measure of the extent to which I failed as a classicist is that I always preferred Latin to Greek.

Before I left, however, I had begun to take pleasure in reading some Latin and even Greek poetry. We had to write verse in both languages ourselves, and this I found totally beyond me; it was a chore for which I simply didn't possess the necessary skills. Translating into English was far less demanding and therefore not regarded as so meritorious. Even the verse passages were only required to be translated into prose, until, that is, I reached the Upper Sixth. It was then that the Headmaster set us the exercise of translating a poem of Catullus into English verse. There were deep murmurings of protest to which he sportingly responded with a promise to write a version himself. When the work was handed back to us he had marked his own at Alpha and mine at Alpha plus. My usual marks tended to be Beta minus and Gamma plus and I have no doubt that this was a gesture of encouragement rather than a deserved reward of merit; but it worked, and incidentally further enhanced my good opinion of the man who had been far-sighted enough to introduce those country expeditions. I set about translating other lyric poems and received further commendations, but they constituted a solitary bright spot in an otherwise undistinguished career of classical scholarship.

Not that it was all wasted, as I hope to show later. All through those years something of the classical way of looking at the world was probably seeping through without my realising it. Put in another way, my building of the world, my own world, would have been very different if I hadn't been made to persevere with the so-called dead languages; but the fact is that the sense of remoteness of academic work from the real world of my own experience, which had so disheartened me at the Glebe House, still persisted. By now, of course, we were no longer restricted to translating sentences which had been specially constructed to illustrate some points of grammar and syntax; we were reading original texts written in the ancient world. Many of my brighter friends found it quite exciting, and I have no doubt that they were helped to acquire their enthusiasm by some excellent teachers, but nobody ever found a way of kindling more than the feeblest spark of enthusiasm in me.

In spite of my tardy progress in academic work my intellectual development was not entirely at a standstill. I was beginning to think for myself. In particular I found myself looking more critically at the very system of which I had become a part. Ideas and assumptions which had been so instilled into me that I accepted them without question I now began to look at more closely. It must have been at about the age of seventeen that I first realised that there is a tendency in most people either to accept or to reject an idea; they are much more reluctant to pull it to pieces, accepting and rejecting selectively its constituent parts. I came to understand that we live in a world of package-deals. Whether in politics, religion, philosophy or any other field of experience, we are invited to accept a syndrome, an integrated system of interrelated propositions, on a take-it-or-leave-it basis.

The danger that one will accept what has always been accepted, and for just that reason, is particularly strong in an English public school with its exaggerated emphasis on tradition. Add to that the tendencies for teachers of the Classics to believe that values which have stood the test of time for two thousand years must be right, and you will see how easy it is to emerge from this sort of educational experience with an inclination to accept a package-deal simply because it seems to have met with general approval. Didn't the *alumni* of the school prove it? The vast majority seemed to have settled for their package-deals after at best the most superficial examination. Most of those who had distinguished themselves did so within the rules laid down by the Establishment. Even the notorious Judge Jeffreys of the Bloody Assize achieved a place in the history books by an over-zealous support of the cause of his monarch. True he died in the Tower of London, but that was because the Establishment itself had changed with the flight of James II and the arrival of Dutch William.

I suppose a very large number of Old Salopians must have become doctors or lawyers, civil servants or businessmen (often in the established family business), clergymen or officers in the armed forces; but not all. Shrewsbury School, in fact, has a long history of occasional rebellion against established ideas. Its iconoclastic heyday came after my time with that band of young men who practised their art in the pages of *The Salopian*, as I had done, before perfecting it in *Private Eye* and paradoxically presenting the nation with one of its establishment institutions. Even that paragon of Conservative virtues, Michael Heseltine, caused a sensation by seizing the Mace in the House of Commons. Later he figured in a resignation, memorable for its melodramatic walk-out, from the Government of Margaret Thatcher, and eventually played the leading role in toppling her from the prime-minister's office.

Long before this Shrewsbury School had produced the odd rebel against accepted habits of thought. Job Orton, the dissenting minister, had practised his non-conformity ecclesiastically in the middle of the eighteenth century, while Samuel Butler, in the second half of the nineteenth, had suggested in *Erewhon* that society could look very different if one took away that framework of accepted values within which it habitually saw itself, and replaced it with others. Even that hero of the Establishment, Sir Philip Sidney, had shown a remarkable independence of mind for a sixteenth-century aristocrat.

But neither *Erewhon* nor *Private Eye* produced more than a ripple compared with the shock-waves which swamped the nation and indeed the world when *The Origin of Species by Natural Selection* appeared in 1859, and of all the innovative Old Salopians the one who really captured my imagination was the below-average schoolboy who had experienced so much difficulty in keeping his mind on his classical studies when all his passions lay in the direction of exploring the world around him with his own eyes.

Charles Robert Darwin had entered the school a hundred and fifteen years before me, but everything I came to know about him seemed to melt that time-span away. I now realise that what I did know at that stage was little enough, but the more I discovered the more I came to see in him the strong points which I admired and the weaknesses with which, from personal experience, I could so well sympathise.

In the first place he had never achieved any academic distinction at school, where he had been accused of being lazy, though he himself protested that he had worked hard. I had some fellow-feeling there!

In the second place - and no doubt this had much to do with his lack of academic success - his principal interests at school lay in subjects which the curriculum offered him no opportunity of studying. J. B. Oldham (1952, 195) tells a story of Darwin that he was once publicly rebuked by his headmaster for wasting time on such a useless subject as chemistry. Shortly after leaving school I was rebuked by mine, albeit privately, for forsaking the Classics at Oxford and taking up a subject which had no business to be in a university curriculum at all!

The third point of contact which I felt with Darwin lay in the belief that the observation of what we could see in the world around us with our own eyes was as important as what we could read in books. To some of my teachers I'm sure this would have seemed an unforgivable heresy. I'm certainly not suggesting that Darwin was uninterested in books; nobody could succeed in formulating an all-embracing theory of life without access to that vast storehouse of accumulated knowledge which is to be found in the literature of the world, past and present. But to some of my contemporaries, including some of my teachers, it seemed that this literature was everything. Unless a famous author had enshrined an idea or observation in some canonical form of words it could scarcely have enough value to warrant the effort of remembering it. Academic learning appeared to be a sort of closed system, the contents of which were there to be passed down from generation to generation, augmented from time to time by the addition of approved new writing but unadulterated by the intrusion of innovative trivia. No doubt this is a gross libel on the academic attitudes of those who attempted to teach me, but the feeling somehow came through that, if I was interested in what *places* looked like, how they differed from each other and why I felt so strongly about them, or some of them, as sources of emotional satisfaction, I was, like Darwin, 'wasting my time'.

The fourth virtue which I admired in Darwin was his courage and forthrightness in communicating revolutionary scientific ideas in a society which was reluctant even to consider them, much less to accept them, if they seemed to challenge the philosophical assumptions to which it had become accustomed. The man who had the stamina and dedication to amass all that information, the inspiration to see in such a complexity of facts, such a simplicity of meaning, the sympathetic understanding to communicate his disturbing ideas in a prose so gentle and restrained as to minimise the trauma which inevitably attends the rejection of long-cherished assumptions yet at the same time the honesty to proclaim a view which was bound to bring him nothing less than severe moral censure, such a man was surely a fit subject for the hero-worship of a none-too-successful schoolboy!

The spirit of revolution in my own breast stirred very slowly. It seems to have begun, as I suppose it begins in most people, in a sense of irritation and dissatisfaction with tiresome trivia long before I went to school. By the time I was into my teens it had started to quicken and to acquire a little more direction, a little more purpose; but it was slow to express itself, partly because a belligerent and aggressive attitude was not a part of my nature, and partly because the very process of questioning helped to establish a concern for that sense of fairness which was one of its own sources of motivation. The counter-argument was always entitled to a fair hearing. So I began to question not only whether the irksome restraints of school life were really necessary but also how they had come into existence. Many of them seemed to have no better justification than an origin in custom, habit and tradition, and I began to realise that tradition is a two-edged weapon. I remember thinking up a little epigram: 'If it's good it needn't be justified by tradition; if it's bad it can't be. So what price tradition?'

At this time also I was rapidly broadening my acquaintance with English literature, but nearly always, where there were visual images of places, they impressed themselves on me at least as much as human actions. The integration of the one with the other always fascinated me. If landscape was a stage for the enactment of events the stage generally dominated. It was a case of 'landscape with figures' rather than 'figures in a landscape'.

It's therefore perhaps not surprising that my growing familiarity with the pleasures of literature, especially poetry, and with the geography of the Welsh Borderland should have paved the way for a particularly cordial response to the work of A. E. Housman whose pre-occupation with Shropshire had ensured for his writing a widespread enthusiasm in its county town, and not least in Shrewsbury School. Housman, of course, was still alive when I arrived there in 1933.

I realise now what may well have escaped me then, that the poetry of *A Shropshire Lad*, to which I found myself so attracted, had much in common with Fitzgerald's translation of Omar Khayyám. There was the same melancholy sense of the transitory nature of human life, the same deep uncertainty of the worth of anything beyond the present, the same ambivalent feeling of wanting to reject an unwelcome religious philosophy while gratefully accepting the pleasure of listening to the sound of the words. Above all there was the same technical structure, the rigid application of an underlying scheme of rhyme and metre of great simplicity and, to me, irresistible charm. That this was an antiquated style, dated, trite even, and by the nineteen thirties already discarded by virtually all contemporary poets, concerned me not at all. It provided for me a passage backwards into the comfort of the preceding centuries of rhyming metrical verse rather than forward into the disturbing disorder of modernism in which I found little either to make sense or to give pleasure.

Since nearly all the places named in *A Shropshire Lad* lie in the southern part of the county, my interest in Housman reinforced the general feeling that Shrewsbury School had a south-westerly aspect. The site itself lay some hundred and fifty feet above the River Severn (Figure 19). It stretched right up to the edge of the steep grass slope on the outside of the meander which all but encircles the town. The

School Buildings, converted from various earlier uses, stood on the very lip and presented quite a dramatic façade towards the river and the town. But, together with more modern buildings of late-Victorian and even post-Victorian date and some rather splendid trees of forest dimensions on the upper part of the bank, they formed an effective visual barrier when viewed from the main part of The Site which consisted of extensive playing-fields surrounded by sporadic school buildings.

All the time, therefore, one was subjected to the feeling that The Site was sealed off on its northern side by a kind of rampart. Beyond lay the town and beyond that the expansive plain of North Shropshire, made familiar enough as a distant feature of the landscape by hours of illicit gazing out of classroom windows, which looked out in that direction, but unattainable in reality by virtue of the school rules which placed all that territory out of bounds. It seemed to add up to a cogent expression of that underlying current of monasticism committed to the renunciation of the world, as represented by the distant prospect, the flesh, as personified in the naughty ladies down by the English Bridge, and the devil as symbolised by the town.

To the south-west, on the other hand, the visible manifestations of confinement were much less apparent. The fence of vertical iron palings was broken here and there by intentionally provided openings, and, in any case, the limits placed on our comings and goings in that direction were enforced, as I have explained, only by constraints of time. Beyond the fence lay the accessible walking-country of lanes and field paths and beyond them the hills of South Shropshire. Here and there nestled the little towns and villages selected by Housman probably as much for the music of their names and their compatibility with the metrical demands of his verse as for the individuality of their personalities.

> Clunton, Clunbury
> Clungunford and Clun
> Are the quietest places
> Under the sun.

I mustn't leave my description of the school site without making brief reference to the residential area of Kingsland which lay immediately outside the School Gates. It had been developed mostly from the time when the Schools had moved out from their downtown site in 1882, the same year in which Darwin had died. The whole of this area, which included some of the school 'houses', had become an extreme expression of the aesthetics of late Victorian living in what might be called the Second Division of the Upper Middle Class. Elegant three-

storey detached houses of red brick, often patterned with bricks of a
darker colour and roofed with red tiles, stood in romantic gardens,
many of them enclosed with brick walls. Tall conifers supported by a
huge variety of lowlier flowering shrubs formed a kind of backcloth to
little lawns fringed with herbaceous boarders and punctured with
weedless beds of annuals dutifully planted out by unobtrusive
gardeners.

The garden of the School House itself, in which I lodged for five
years, was one such. It was bounded by a stone wall fronted by
shrubs, a sort of soft lining to a hard shell, and, since my housemaster,
J. R. Hope-Simpson, was a horticulturist of some competence, it was
kept in immaculate condition. We had no right of access to it, but he
made liberal concessions to any of us who showed enough interest to
want to wander around in it or even sit there and revise for
examinations. My own interest was not so much in the plants as in the
atmosphere which, late on a summer evening in this calm, scented
retreat, often drew me to savour a sensation which I found quite
enchanting, though I would not have been able at that stage to suggest
a rational explanation.

It may have been at Shrewsbury that I first became aware of
what one might call a typology of landscape, based not so much on
what places looked like as what they *felt* like, though it was clear
enough that the appearance and the feelings were interdependent.
The phenomenon was well exemplified, in the immediate environs of
the school, by three kinds of ambience, all contiguous, all inter-
related yet all strikingly different. First there was the open expanse
of the playing-fields; secondly there was the dramatic bend of the
Severn, flanked on the outside, 'our' side, of the meander by steep
cattle-grazed pastures with irregular tall trees, and on the inside by
the smoother mown lawns and the still superb lime avenue of
Shrewsbury's splendid public park, The Quarry; and thirdly there
were the enclosed gardens, little nests of elegant Victorian privacy
from which rose gables, chimneys and coniferous spires into that
ubiquitous symbol of universal freedom - the sky.

Aedes Christi

In October, 1938, I was admitted as a commoner, that is to say not a
'scholar', to Christ Church, Oxford. Not Christ Church College; there
was no such place. It was, of course, a college, but with a small 'c'.
You might also suppose that on admission I had become a student,
but again you would be wrong. 'Students' in Christ Church are
dons; I was an undergraduate.

Whether this was the right college for me I wasn't sure. *Aedes
Christi*, colloquially known as the House, had the reputation of being

FIGURE 19. PART OF THE SCHOOLS, SHREWSBURY. View towards the north. The School Buildings (with central clock tower) are flanked by the chapel (left) and the School House (right) which collectively form a barrier separating the area referred to in the text as 'The Site' (foreground) from the River Severn beyond and approximately 150 feet lower. The public open space known as 'The Quarry' lies beyond the river. Photo by Aerofilms, July 1969.

FIGURE 20. 'THE HOUSE' FROM THE SOUTH. The street on the left is St Aldate's, Oxford, and the buildings of Christ Church lie to the right (east) of it. The largest quad is Tom Quad and the spire marks the position of the Norman Cathedral which is also the college chapel. The large building in the right foreground of Tom Quad is the Great Hall and the first-floor windows immediately adjacent on the left are those of my rooms (1938-39). The trees in the bottom right-hand corner of the picture indicate the north-west corner of The Meadows. Photo by Aerofilms, June 1968.

the most aristocratic of the Oxford colleges, and the list of titled young men among my contemporaries probably bore this out. It also nurtured many of the country's leaders, not least in affairs of state. In the postwar period alone it has provided a Lord Chancellor, a Chancellor of the Exchequer and no less than three Foreign Secretaries, two of whom went on to become Prime Ministers. True, it spanned a remarkably broad spectrum of social classes but heavily weighted towards the upper end, and, if one's financial resources, like mine, were no more than adequate to make ends meet at a fairly frugal level of living, one had to make it clear from the outset that there were many popular social activities in which one didn't wish to be included, just in case one aroused the expectation of repaying some act of hospitality or conviviality which was beyond one's means.

I therefore found myself turning to less expensive forms of entertainment which didn't appeal to most of my friends, and these included the exploration of the surrounding countryside on a bicycle as well as spending a good deal of time walking around the City of Oxford on foot without having any particular destination in view. Many of my contemporaries no doubt found this hard to understand, so I often undertook these expeditions alone. If it's true that like-mindedness is conducive to congenial relationships, one can hardly be more like-minded than with oneself. One of the things that Oxford taught me, therefore, although there was evidence of it much earlier, for instance when I went on my 'railwaying' expeditions at Shrewsbury, was to be content with my own company. Certainly I had many friends, and I saw much of them, but they were not indispensable for any particular occasion. I would never have said, for instance, that I couldn't go to a play or a cinema simply because I had nobody to go with. Lack of money might be, and often was, a constraint, but not lack of company. If it was available, so much the better; if not, I would go alone.

My immediate environs in college were very different from anything I had previously experienced (Figure 20). I was allocated rooms on the first floor on the south side of the main quadrangle, Tom Quad, so called after Tom Tower which accommodated the largest bell, Great Tom. Here most of the accommodation was non-residential, housing such facilities as the Junior Common Room and the Steward's Office, or was reserved for dons. Much of the east side was occupied by the Deanery. Most undergraduates had a study and a bedroom, but there was one suite of rooms 'in Tom' which had for long been allocated to one of the dons ('Students') but had recently become available for undergraduates, and, since it had two bedrooms, it had to be shared. Fortunately I found my room-mate congenial. By his early teens he had reached a height of six foot seven, though by the time I met him he had acquired, perhaps even cultivated, a slight stoop and had lost an inch. He was both a brilliant quarter-miler and a very competent pianist, and it was for this last reason that we had a grand piano in our rooms. The association with Guy Wethered proved a happy one, not least because I came to know his circle of Old Radleian friends as well.

I was told that the rooms we occupied (Figure 20) had been occupied by Albert Einstein when he had been a Visiting Professor in Oxford in 1931, though I never took the trouble to check this out. Not every staircase in those days had a bathroom. The process of equipping a Tudor building with 'mod. con.' takes time, and by the 'thirties improvements had resulted in the provision of facilities on about every third staircase. We had a loo, of course, on our staircase, but to take a bath one had to go out into the quad and re-enter by another door. Apparently the sight of the famous man, already a caricaturist's dream, trotting along with soap and towel, had understandably been a bit of a crowd-puller, and to avoid embarrassment Einstein's 'scout', or manservant, had acquired a very large galvanised iron bath which he used to place in front of the open fire and fill with a few buckets of warm water. The scout, Hicks, who could well have been the prototype for every film director's image of the English butler, was still serving the same staircase when I arrived a few years later, and the bath still hung from a nail on the wall outside our room. More than once I walked into it while fumbling my way up the staircase in the dark. After one such misadventure, which caused me little damage but made a formidable noise, I was indiscreet enough to suggest that perhaps it might be moved. It was then that I learnt its history. I also got the message that, just as there is a proper place for galvanised iron baths which have become historical monuments, so also for undergraduates, and I never made that mistake again.

Of all the rooms I ever lived in this one came closest to fulfilling the concept of the castle. To the south the large living room looked out over the gardens towards the Thames. To the north my bedroom looked out into the quad. The buildings were essentially Tudor and they completely surrounded the open square, leaving only a few points of access here and there. There were other quadrangles in the college, all inter-connected by a system of passageways, cloisters and staircases to which access from the outside world could only be obtained through one of the three main gates. Visually it was much more enclosed than either Shrewsbury School or the Glebe House, but because ingress and egress were relatively unrestricted, at least up to 12.20 a.m. (after that there were severe problems), it felt much more like that supreme image of a personal retreat, Stibbard Rectory. It was *I* who chose when to go out and when to come in, and, more importantly, where to go in the meantime. It was essentially a sanctuary rather than a prison.

In short, living in Tom Quad, one could easily imagine what it must have been like to live in a proper castle. My cosy little bedroom, encased in stone walls, looked out onto an extensive but confined space, also enclosed in stone walls. We went to bed fortified by the knowledge that unauthorised aliens would be excluded from it. Its sanctuary was the monopoly of a privileged *élite* of which I had become a member, and the whole apparatus of seclusion was guarded by the night porters while we slept the sleep, if not of the just, at least of the secure.

My recollections of Christ Church contain a host of visual images, many of them particular, like the curious ogee shape of the top of Tom Tower and the portraits of the long-deceased statesmen and divines which graced the walls of the Great Hall; but for me the visual personality of the college can be summed up in two impressions of a more general kind. The first is of the afternoon sunlight creaming up the limestone and infusing it with a warmth which was almost akin to life. The second is the very opposite - deep black recesses leading out of the quadrangles like rabbit holes in the boundary bank of a field. Some of them were regular and formal, like the arches which invited access to the Normal chapel, dignified by the episcopal chair as the cathedral church of the Oxford Diocese, or the large square hole, which is how one may best describe it, forming the entrance to the Great Staircase. Others were irregular and informal - not much more than nooks and crannies - and they might lead almost anywhere. Hundreds of little rooms, and some not so little, nestled away round blind corners and up creaking stairs. Most of these recesses were furnished with electric lamps which the Steward had ensured were of the lowest power available on the market, and these sufficed to dispel some of the gloom, but the overall impression was one of blackness, absolute and unrelieved.

The romantic effect was achieved neither by the sunlit stone nor by the shady recesses but by the immediate juxtaposition of the one with the other. Together they seemed to symbolise a kind of dualism, not in this case the traditional dualism of light and darkness as symbolising good and evil - there was nothing evil about this kind of darkness which in its own way was friendly and inviting - but some as yet undefined dualism between two concepts which only began to make rational sense to me when I gave the matter some serious thought many years later.

An important part of the environment of Christ Church was the expanse of open space which separated the college from the River Thames rather more than a quarter of a mile away. It was known as Christ Church Meadows, invariably abridged to 'The Meadows'.

The meadows themselves, cattle-grazing pastures, were fenced off from the several gravel paths to which the public had access and along which I took a daily stroll, usually alone, down to the river. Tall trees flanked the paths in straight avenues, while other, informal groups rose from the damper parts of the adjoining levels. They represented respectively nature tamed and untamed and gave me, I believe, my first really pleasurable experience of flat land. My early recollections of the Fens had prejudiced me against flat places, but the Fens were almost wholly arable, almost treeless and vastly more extensive. Christchurch Meadows by comparison were green. They had a deceptive appearance of being wholly natural, and the middle-distance views, terminated by tall screens of poplar, willow and other species, were interrupted here and there by closer clumps of quite thick vegetation. The Thames itself fitted into this *ensemble* and gave it a kind of heart, though I personally preferred the more intimate waters of the Cherwell which, like all Oxonians, I explored by the leisurely punt.

Under the regulations which then prevailed one was not allowed to proceed to read for an Honours Degree until one had surmounted a preliminary hurdle in the form of the First Public Examination. People with high academic ambitions took Honour Moderations usually after four or five terms, but those with lesser aspirations, like myself, could take 'Pass Mods' as soon as they felt able to submit themselves for examination, so I set about getting this out of the way as soon as possible by taking an unlikely mixture of subjects, Latin and Greek, most of which consisted of set books which I had already read at school, Ancient History, and English Constitutional Law and its History. The last was entirely new to me and I chose it partly because, for some reason which was never explained to me, it exempted me from one set book in each of the languages. This was not, I admit, the worthiest motive for the choice. Since I passed in all these subjects at the end of my first term, it follows that the standard was not too exacting.

I had gone up to Oxford with no idea of what I wanted to read for my Finals, and I used this first term shopping around. My continuing contact with the Classics confirmed my feeling that it was time for a change, and after considering various other subjects I was accepted to begin the Honours Degree Course in Geography the following term. It was a decision I never regretted.

My tutor when I entered Christ Church was Robert Mortimer, later to be Bishop of Exeter, but when I moved to the School of Geography I was deemed to be beyond his help. Not that any other tutor could be found in The House at that time who had any knowledge of geography. So for a 'moral tutor' I was assigned to Dr A. S. Russell, a humane Scottish scientist whom I liked, while for academic tuition I was sent to Freddie Martin (whom I also liked), then a don at Oriel, later a Fellow of St John's, and who became, and remained, a good friend until his death some forty years later. Freddie was only a few years older than I, and, though an able scholar and a good teacher, he did not project the image of what I would then have thought of as an 'academic figure', a term I would still have regarded with profound misgiving. His tutorials were intellectually demanding and he communicated to me an infectious interest in his subject, but the over-riding impression he presented as one entered his room was one of beaming geniality. Had it fallen to my lot to be initiated into my new subject by a more formal and forbidding personality my conversion might well have been less enthusiastic.

Paradoxically I believe the very fact that I hadn't studied geography at school, at least since the age of twelve, was one of the points which commended it to me. Even the most superficial glance at the syllabus suggested that it was possible to study academically the sort of things to which my mind had habitually turned when I was supposed to be concentrating on other matters. The subject already had a 'good image' in my own estimation, though a much more dubious one in the eyes of most people who were moved, like my headmaster, to venture an opinion. It was variously regarded as a soft option, a gimmick or just a joke in comparison with the more long-established disciplines. History, for instance, was a highly respected subject, almost in the same club as 'Greats' or Divinity, and certainly Modern Languages. The pure sciences had also attained a measure of acceptability (except among the staunchest Classicists, of whom I seemed to know quite a large number), though perhaps an acceptability of a somewhat different kind. But Geography!

I was not at all worried by this attitude, partly, I suppose, because I knew it was born of a gross ignorance of what the subject actually comprised. Some of my friends had actually done geography at school and based their assessment of what a university course must be on that. If only they knew! But they didn't even want to know that there was a whole *rationale* behind the evolution of the hills, the rivers, the coasts, the farms, the cities and the wild places of the world. They seemed not even to be potentially interested in such things.

In addition to the lecture courses on geography proper we were encouraged to attend others on related subjects. So I found myself listening to lectures on astronomy by an American professor whose name I have long forgotten, and on various branches of geology. One incident during my first term in the School of Geography I remember with great clarity. We had just been introduced to the Carboniferous

Limestone with black-and-white slides of the landscapes associated with this rock in Derbyshire, Mendip and the Wye Gorge. As the lecture proceeded I began to grasp at something pigeon-holed in the deeper recesses of the memory and when it ended I went at speed to the map library and asked for the geological map of the Isle of Man; and there, sure enough, between Castletown and Port Erin was an outcrop of the Carboniferous Limestone. I knew before I saw it what I was going to find, so clear had been the image retained in the mind's eye from that brief visit at the age of six. Even though much of it is covered with glacial deposits there must have been enough of the drift-free limestone to communicate a distinctive 'feel' which I was able to recognise in the lecturer's slides, even without the aid of colour.

Every Saturday we used to have an excursion into the surrounding country. These excursions were conducted by lecturers from the School and they took us into a fair range of landscape types from the Cotswold Edge in the west to the Chilterns in the east. Rural scenes of a kind which had long been familiar as objects of attraction were now being presented to me as the subject matter of academic enquiry. It was a revolutionary experience and it held out a prospect of quite a new way of looking at the world. Side by side with this academic approach to the study of the South Midlands, I continued to fit my newly acquired information into the sort of non-rational, emotional system of values which continued to influence the way I not only thought about, but also felt about places. The whole of this tract of country consists of a series of escarpments where the limestones and other harder rocks come to the surface, separated by clay vales, all trending in a north-east to south-west direction, and this alternation of ridges of high ground, with fine extensive views, and troughs of low ground, which, though denied the long views, were compensated by a rich sense of shelter, provided a kind of matrix into which the towns and villages of the region were set. Oxford, for instance, had a feeling of openness along the valley of the Isis-Thames to the south and the north-west, which contrasted strongly with a sense of enclosure deriving from Headington Hill and Boar's Hill to the east and west respectively.

I was fully conscious that this feeling would hardly have been comprehensible to most of my friends, not least because many of them wouldn't have known where to find north and south anyway. By the same token many of the things about Oxford which excited them were of no interest to me. I well remember some of the more affluent of them setting off in tails and top hats for the Royal Enclosure at Ascot and feeling almost arrogantly self-congratulatory that I had more worthwhile things to do. Put in another way, I was already deeply immersed in prejudice, but it didn't strike me like that at the time. Not that all Members of The House fitted the Ascot image. There was among them quite a strong sub-current of left-wing political opinion and the Conservative Party didn't have matters all its own way. Two young dons who had rooms on the next staircase in Tom Quad, Patrick Gordon-Walker and The Hon. Frank Packenham, later Long Longford, were destined to become ministers in Labour Governments; but the politically successful *alumni* of the college swung the balance heavily to the Tory side. Nor was this allegiance confined to the top of the college hierarchy. When Lord Hailsham, then the Hon. Quintin Hogg, and himself a Member of The House, was elected as the Conservative Member for Oxford in a famous by-election, I well remember Hicks coming in to wake me the following morning with the announcement, 'Sir, Oxford, nay England, is saved!'

Before we leave Oxford there is one little episode I should tell you about because in retrospect I suspect it was probably my first act of public social protest. A representative of the Oxford Union Society called on me with a view to persuading me to join it. I was not particularly politically minded then or at any other time, but I was generally interested in current affairs and I found myself attracted by the idea of hearing some of the best orators and debaters in the country, if not in the world. My visitor, however, confirmed what I already knew, that women were not eligible for membership, and he left without my subscription. Somehow the Oxford Union managed to survive without me as also without any of those far more gifted and politically active members of Somerville, St Hugh's, Lady Margaret Hall and the other women's colleges. Needless to say this gesture of protest went unnoticed, but it served for me as a kind of initiation and a precedent when, not many months later, I was called on to make a protest of a different and more far-reaching kind. But that's a story which must wait till the next chapter, because we still have some filling-in to do.

Teenager at Large

The years between the ages of nine and twenty were not spent wholly in academic institutions. The holidays and vacations afforded opportunities for acquiring new material for making that private world of my own. Radical as were the changes which overwhelmed me from the day I first went to the Glebe House, they only affected my life during term, and for a quarter of the year, or maybe slightly more, I was able to continue the sort of life-style that I associated with home and family. Three times a year I returned to that little piece of North Norfolk which had come to satisfy in a unique way my needs for territorial assurance. Stibbard, thank God, was still there, and, as absence makes the heart grow fonder, so it became even more precious a possession than before.

Let me begin with a few words about those expanding horizons of a more limited kind within reach of the home base. The acquisition of a bicycle before I went to school had begun to open up new pastures, and during my teens I availed myself of many opportunities to explore the by-ways of the area around Fakenham. As my range increased it became practicable to reach the little villages along the North Norfolk coast. The nearest place was Wells, and there were many different ways of getting there. The most direct was through those curiously named villages, Great and Little Snoring, which one might think were the creation of some comedy script-writer, like Much Binding-in-the-Marsh, but are actually authentic place names, like Knotty Ash. An acceptable alternative was the Sober Road, so called because it passed no pubs between Fakenham and Wells. Each route offered its own mood; the former, which largely followed the shallow valley of the little River Stiffkey, was more enclosed, more sheltered, and particularly where it passed through the villages, more cosy. The latter followed closely a ridge of higher ground and commanded wider views particularly to the east, and, in a county where there are not too many high vantage-points affording extensive views, I found that exciting.

On one of these teenage cycle rides I was exploring the country which lay to the west of the Sober Road when I was attracted by a large belt of tall trees. In general this north-western corner of Norfolk is not as richly endowed with woodland as the area to the south-east of Stibbard, that is to say in the direction of Norwich, so I turned aside to investigate. The belt of trees appeared to be a screen against the wind, not unlike other shelter-belts I had seen often enough, but it was much bigger, measuring its length in miles rather than hundreds of yards. I left my bicycle in the ditch and walked through the wood until I emerged at the other side on the lip of a kind of saucer. Unlike all the country around, which consisted of hedged fields, mostly arable but some under pasture, it contained a great open sward studded with giant trees spreading their canopies over little oases of shade. I was overcome by a compulsion to stand and look at it, a feeling which was not rational but emotional, like falling in love. It just didn't belong to that North Norfolk which I thought I knew; if I had ever seen anything like it I suppose it had been in picture books.

What had happened was simply this; these little lanes and by-ways had brought me to the back of Holkham Park and I was looking at the legacy of the imagination of William Kent who had conceived this idyllic landscape nearly a couple of hundred years before. Kent I had never heard of, nor was it a name I was to hear for many years more. Indeed in that ecstatic moment I probably gave no thought to the fact that this enchanting landscape must have been devised by some person or persons and artificially created out of its agricultural environs for the simple purpose of arousing the sort of response it had just awoken in me.

One of my interests which blossomed in my early teens was fishing. By the time I was twenty I had virtually given it up, because the excitement of proving one's superiority over the craft and self-protective cunning in animals was outweighed by the distaste of killing or merely injuring them; but while the vogue lasted it took me into some landscape types which I should probably not otherwise have frequented and which have left behind indelible visual impressions.

The most accessible places for fishing lay along the course of the River Wensum. The nearest point was at Great Ryburgh (Figure 9), less than a couple of miles from home. There I occasionally fished in the river itself, but quite often in the so-called 'Main Drain' which flanked it and carried off surplus water from the adjacent meadows. In the Main Drain I used to catch crayfish and once even a pike. I would lie on my tummy on the bridge itself, not much more than a culvert, my head and shoulders pushed under the railings, and aim to tempt the fish underneath. It was not a very comfortable position and I was frequently interrupted by traffic on the road and occasionally by passers-by who asked questions I didn't really want to answer, such as whether I had caught anything, when they could see perfectly well that I had not.

Now and again I rang the changes and bought a permit to fish further downstream at Guist, pronounced like the German *Geist*. Here the river wandered through woods and meadows well away from the road. I remember once sitting on a tree-trunk which spanned it, presumably where it had fallen, and as I watched my float drift slowly down the stream I was startled by a large wet object hurling itself onto this fortuitous bridge at vast speed and pulling up a couple of feet from my face. In fact I don't suppose even a large otter is so very big, but at that range it looked enormous. I remember thinking that I had somehow intruded into its home, the place where it naturally belonged, and that I was interfering with its legitimate coming and going. Suddenly I was able to see that short stretch of slowly moving water, fringed by the reed beds, as a kind of stage where this beautiful creature acted out its life story, a stage so different from the one where I played out my own life, yet entirely comprehensible even to a non-otter as embodying the essence of the concept of 'home'. It had hiding-places, feeding-places, playing-places; it had little secret channels for coming and going through the reed beds; the place and the animal obviously belonged to each other in the most intimate way, just like me and Stibbard Rectory.

The bridges over the Wensum at Ryburgh and Guist are about three miles apart, and between them the river flows lethargically through the grounds of Sennowe Hall. Within the park an artificial lake had been excavated from the floodplain and stocked with fish, and this was another venue where I used occasionally to go for roach and other coarse fish, dangling my line in the shallow water from the balcony of the little boat-house (Figure 21). Although it was separated from the river by only a few feet, the lake aroused feelings of quite a different kind. In a way it was more open, in another way more enclosed. There was a thick bank of trees which screened off the river on the south side; on the north it merged into the grassy sward of the park, reminiscent in a way of the park at Holkham, yet in a way so different. The park at Holkham was a deliberately contrived aesthetic invention of the eighteenth century, a pleasure-ground made within the idiom of its own time, just as the gardens and gabled houses of Kingsland outside the school gates at Shrewsbury were expressive of late Victorian taste. Here at Sennowe I had discovered a park whose function was like that of the park at Holkham but whose artistic idiom was nearer to that of the Kingsland Victorians, though the Hall itself was, strictly speaking, an Edwardian building. So there was a splendid avenue, straight as a ruler, reviving a seventeenth-century style, which led into a more open park, reminiscent of an eighteenth-century idiom but with a liberal sprinkling of those huge conifers, which the Victorians loved, and a pointed clock-tower apparently mimicking the surrounding coniferous spires (Figure 22). The whole atmosphere of the place aroused in me that 'being-in-love-with-a-place' feeling which was by now becoming quite a familiar experience.

The death of my grandmother in 1932, when I was twelve, broke the pattern of annual visits to Bradley Hall. It may have eased somewhat the financial anxieties of my father, though it did not remove them altogether. He and my mother made it a priority to find school fees for myself and my sister, and, education apart, there was a general acceptance that we never spent more money than was necessary to reach whatever objectives we had set for ourselves. I suspect that, reading through this book, there may be occasions when you will detect signs of an underlying puritanism in my attitude to the spending of money, but it isn't as simple as that. Puritanism implies some virtue in the avoidance of pleasures; the thrift practised by our whole family was quite different in that it involved saving the money we might have spent on a lesser source of pleasure so that it would be available to spend on a greater.

So it was that the principle of obtaining the best value for money commended camping as the solution to the problem of what to do for a summer holiday. My father bought a ridge-pole tent

FIGURE 21. THE LAKE AND BOATHOUSE, SENNOWE PARK, NORFOLK. View to the south-east. Photo by the author, 1986.

FIGURE 22. THE SENNOWE SKYLINE. Note the juxtaposition of open and enclosed space and the variety of contrasting shapes contributed to the skyline by the combination of deciduous and coniferous trees and the dramatic architecture of the tower. View to the north-west. Photo by the author, 1986.

made by Blacks of Greenock as part of a large consignment ordered by the Icelandic Government for the millennial celebrations of the first Icelandic parliament in A.D.930 and sold off shortly afterwards at a bargain price. It bore in one corner the signature of, I suppose, the Scottish seamstress who had been responsible for its manufacture and it was always known as the Bessie Wilson. By modern standards it was an impossibly cumbersome affair and there was no way we could have fitted it with all the other luggage into the car; so my handyman father set about making a small luggage trailer to carry all our requirements, and this meant that we were able to enjoy many comforts, like folding chairs and tables, which my mother was glad to have, though I'm pretty sure my father, as well as my sister and I, would have been quite happy to manage without.

Thus equipped we set out *en famille* for more distant parts, and over the next few years we reached Land's End and John o'Groats in successive years, North Wales and other less remote destinations. As with the journey to Bradley the process of getting there was by no means the least pleasurable part of the experience. The roads at that time were, by comparison, with the present, primitive but empty. The Great North Road, for instance, which we joined at Newark, went right through the middle of all the towns which it had been constructed to link up as a turnpike road in the seventeenth and eighteenth centuries - Retford, Bawtry, Doncaster and the little Yorkshire coaching towns of Wetherby, Boroughbridge and Catterick. Further north in Newcastle it combined the function of the principal road link between the English and Scottish capitals with that of one of the city's principal shopping streets. It still consisted largely of a single carriageway with one lane for each direction. Progress therefore was pretty slow, and, with a thirty-mile-an-hour speed restriction in force for trailers, three days had to be allowed to reach the Highlands, which, needless to say, made the destination seem even more remote and its attainment correspondingly more arduous.

Commercially developed camping grounds in the middle 'thirties, except in a few popular coastal locations, were few and far between. They tended to be crowded and regimented and were anathema to my father whose antipathy was easily communicated to the rest of us. True, they had toilets of a sort, but we preferred to dig our own hole in the ground which we duly enclosed in another tent of Punch-and-Judy shape. The pastime of camping, however, was so little institutionalised and so little constrained by regulation that one had every expectation of finding a suitable field almost anywhere. The A.A. provided its members with lists of farms which they had inspected, but many farmers all over the country just hung up notices and charged half a crown a night, sometimes less, sometimes nothing. Others seemed quite content to give permission with no more formal introduction than a knock on the door and a polite request. Sometimes this meant sharing a meadow with its permanent tenants, and one soon learnt that cows are the most inquisitive of creatures and also to take care where to put one's feet. The choice of a site for pitching the tent might also have to be determined by the spaces between the cowpats.

In these circumstances we found some truly delightful sites. It was a wonderful sensation to get the tent pitched just before darkness fell, to cook an evening meal, troubled only by the ubiquitous midges, and finally to snuggle down in the blankets and feel that one had been at least partially united again with that same Mother Nature from whom civilisation had so relentlessly attempted to set one apart. Outside the tent were the smells of the countryside and the noises of the night. Inside were security, cosiness and lively young imaginations trying at one and the same time to anticipate the adventures of the morrow and to get to sleep.

My mother's increasing desire for just a few more creature comforts as she reached her middle fifties, coupled with my father's ambition to press his inventive ingenuity a stage further, led to the luggage trailer being supplanted by a caravan. At Easter, 1936, my parents came to fetch me home for the holidays from Shrewsbury. Usually I went home by train, but on this occasion there was an exhibition of caravans at Wolverhampton so they decided to kill two birds with one stone. My father, who on occasions like this discarded his clerical collar, made the most of it. He examined one caravan after another, inside and outside and, most important, underneath. By the time we got back to Stibbard he had his dream caravan practically designed. During that summer he worked assiduously. He visited several scrap heaps and old car dumps, fitted together the main frames of two cars and had the garage attach the springs, this being the only job he had done professionally. I helped him with the carpentry during the school holidays. My mother and Helen busied themselves with the upholstery and the soft furnishings, and in August, 1936, we were ready to set out for North Wales. The whole job had cost just thirty-two pounds.

That, then, was how I discovered the farthest extremities of Great Britain. We were a very closely-knit family (Figure 23), and this pattern of holidays persisted right up to the outbreak of war (Figure 24). The days were given over to a variety of occupations. We had our full share of the usual activities of youth and I suppose the beach often had the first claim on our time, but increasingly exploration for its own sake tended to occupy a major place in our programmes. We would usually make for the wilder places and explore them on foot as well as in the car, invariably with a map, fitting what we discovered into our emerging images of each distinctive locality.

FIGURE 23. 'A VERY CLOSELY-KNIT FAMILY'. My father, Canon J. A. Appleton, my sister Helen (now Mrs W. G. Cook), myself and my mother, Lilian Appleton, 1938.

FIGURE 24. 'THIS PATTERN OF HOLIDAYS PERSISTED . . .' A 'Vest-pocket Kodak' snapshot taken near Onich, Loch Linnhe, Scotland, in 1937. View to the north. My mother prepares a meal on the Primus stove as my father looks on, proud, no doubt, of his handiwork, the caravan. The guy-ropes of the Bessie Wilson (left) serve as a clothes line. Photo by the author.

It was during those summers from 1933 to 1939 that my early predilections for the mountains of the Lake District blossomed into a kind of ideal image. A decade or two later the founding fathers of England's National Parks were to encounter criticism from some quarters for having one-track minds, for only being able to see beauty in the wide open spaces of the uplands and for being insensitive to the charms of all other types of landscape. I believe at that time I probably came pretty close to their thinking. There were no landscapes I had yet encountered at first hand that could compare with these.

Since the summits of North Wales were marginally higher than those of the Lake District, and those of the Scottish Highlands were marginally higher still, and since the open spaces between them were greater and emptier in Scotland than further south, there became established a kind of hierarchy of excellence. The Scottish Highlands took the Gold, North Wales the Silver and the Lake District the Bronze. There were many other places I came to love dearly, the Wye at Symonds Yat, the Pennines, the Southern Uplands, innumerable stretches of the coastline and pleasant rural retreats of all sorts, but their excellence seemed to lie in the extent to which they mirrored at least some of the characteristics of that sublime country north-west of the Highland Boundary Fault. I should be embarrassed if I had to defend such a simplistic classification today, but I have the clearest recollection of how I felt about natural scenic beauty in 1939 and perhaps for some little time thereafter.

The perception of distance had become perhaps the most exciting of all landscape experiences. Those early exercises in identifying the several peaks of the Lake District, as seen from each other, and placing them within a geometrical framework, by which their geographical relationships could be understood, were now seen as the preamble to more ambitious exercises of the same kind. The scale of Scotland was different from that of the Lake District, and if the visibility of distant mountains was a source of delight, the greater the distance the greater the satisfaction. The view from Ben Nevis, being the highest point in Scotland, was a memorable one, but I think one of my clearest recollections is of climbing Ben Wyvis, further north, because, although it isn't so high, it stands in splendid isolation from all its neighbours and the distant heights in all directions are perceived across an intervening expanse of plain.

There was one destination, however, attained a number of times during those years, which was no less magnetic in its own very different way. We lived only about a hundred and twenty miles from London, though this seemed a more formidable distance to travel in the nineteen thirties than it would today. We used to stay with friends in Catford and from there we were taken to see the various sights. As we progressed into our teens the theatre was added to the schedule of events. But it was not so much the things we could see and do in London, exciting as they were, which fascinated me as much as the place itself - the streets, the squares, the feeling of confinement between great blocks of masonry from which a blessed relief was afforded by that great linear feature which could not be built over. The River Thames provided a sense of open space which split the Metropolis in two and invited a variation on Kipling's lines that all Londoners seem to live by, even if they don't admit to believing it:

> Oh, North is North, and South is South,
> And never the twain shall meet . . .

There are many odd incidents connected with London that remain in the memory. I will tell you about just one. It was a wet night and we were going out by train from Charing Cross to Hither Green when I saw the word 'WAR' in huge red letters in the sky. The vision was accompanied by flashes of brilliant light as if the whole place were under bombardment, and for a moment I fancied that it could almost be some celestial portent of a disaster to come. It lasted only a few seconds, but this was long enough to stir the imagination into constructing a picture of horror and devastation which I recollected vividly enough a few years later when I was blown off my feet by a land mine in the Camden Road.

The explanation of the phenomenon was simple enough and involved no appeal to the supernatural. A certain firm of whisky distillers had placed their name in neon lights on a tall building, but on that particular night the first two letters of 'Dewar' were either not functioning or were concealed by some intervening object. As for the flashes of light, I hadn't yet learnt that, in wet weather, the third-rail system of the Southern Electric Railway invariably produced arcing on a grand scale.

It may not have escaped your notice that my experience of landscape up to this time had been entirely confined to Britain. When I went up to Oxford I was probably more widely travelled within its bounds than almost anyone I knew of my own generation. I had visited every county in England, except Sussex, and most counties in Wales and Scotland, but although my interest in foreign parts had been aroused at an early age, the world overseas was for me a kind of alternative system, like the Kingdom of Heaven, partially described but as yet unperceived, a world parallel to that of my own experience. From various sources I had built up innumerable visual images which had been carefully associated with particular places. In the cinema such pictures had come even more closely into line with the reality of my own experience, even though the moviemakers, in converting them into moving images, had as yet been unable to convert them also into colours. But it was still an unreal world, a kind of shadow of the one in which I had been an involved, participating member.

Many of the people I had come to know at school had travelled quite extensively on the Continent and some further afield still. Some of them were the sons of expatriate English parents who went 'home' to exotic places in the school holidays. At Oxford I numbered among my friends a New Yorker, a Sydneysider, and a South African and all this had led to a powerful desire to break through the barrier and cross the English Channel. Early in 1939 the opportunity arose for me to join a party of undergraduates who were loosely affiliated by a common interest in religion and were arranging to visit some places of particular significance in France and Belgium. Unfortunately a planned visit to meet Karl Barth in Switzerland fell through, but the rest of the programme was undertaken in March 1939. I had to let the University Fives Captain know that I would not be available if selected to play for the University, this being the only occasion on which I ever came near to a Half-Blue or indeed distinguishing myself in any sport, but it was a small price to pay.

We crossed from Newhaven (which enabled me to strike Sussex off my list of unvisited English counties at the last minute), and set foot on foreign soil at Dieppe. To many young people today travel is such a commonplace experience that they might find difficulty in understanding the sense of occasion which this represented for me. Had I not been interested in what the overseas world looked like it would, no doubt, have passed less dramatically; but I had been passionately interested. True, the world of Greece and Rome, whose geography had caught my imagination more than its literature, lay well beyond the territory I was now about to explore, while the more colourful lands of the missionary exhibitions were incomparably more remote, but it was a start.

The discovery that, despite all the early evidence to the contrary, France was not a country of various shades of sepia but that it blazed with real colours in 3-D was no less exciting because reason had long ago told me it must be so. Not only the sights but the sounds and smells also, quaintly different from their English equivalents (those eunuch steam locomotives with their falsetto whistles!), were suddenly real and not merely perceived, as Saint Paul says, through a glass, darkly. It was as though some devout monk, spending one lifetime in preparation for the next, and knowing only roughly what to expect, had suddenly been given a return ticket to Heaven to bring immediately within the grasp of his senses those visions which his faith had led him to believe he would one day see, but never with mortal eye.

Our first visit was to the Seminary of the Russian Orthodox Church, which, not being encouraged in the Soviet Union, had settled down in Paris. Our meetings at the Seminary were no doubt the principal source of interest to the theological students, and I was by no means uninterested myself; but to me other impressions were more vivid and more memorable. Certainly the visit to the Russian Orthodox Cathedral was one of the highlights, largely because the music had a quality which I had never encountered before and which made an immediate appeal in that dark, cavernous candlelight. I remember also an old men's hostel in which the atmosphere, in the literal sense, was scarcely believable. There must have been dozens gathered together in quite a small room, huddled round tables, sprawled on bunks or on the floor, and all, I swear *all*, smoking something - shag, opium, who knows? It was like a gallery of theatrical masks bedded in a matrix of floating smoke and faintly illuminated by a feeble yellow glow. I have no idea who they were or what they were doing there, but the picture is as clear now as it was the following day.

For me the programme itself was of secondary importance. The main thing was that it enabled me to go to Paris at a price I could afford. What really excited me was the prospect of turning into reality that scrap-book city which I had put together out of the pages of *Ma Première Visite à Paris*, at the Glebe House. The discovery that Sacré Coeur, The Eiffel Tower and the Champs Elysée all had exact geographical locations and maintained particular and unchanging relationships with each other, just like the peaks in the Lake District, was no less exciting because I already knew it. It was as if a wall full of family portraits suddenly became endowed with breath and with movement and turned into a roomful of real people.

From Paris we went to visit a Lutheran community in Strasbourg. Here again, while I have a vivid recollection of the train journey as we passed through the eastern escarpments of the Paris Basin by Bar-le-Duc and Nancy, and of the city of Strasbourg itself, I can remember next to nothing of the people we visited. Two incidents particularly stick in my mind. The first was a visit to the monastery of Sainte Odile high up in the Vosges. Never mind the theology, the view across the Rhine Rift Valley eastward to Germany was a revelation. Distance, conifers and a dash of snow made sure it would not be forgotten, while the curious Alsatian architecture and a real stork's nest on a chimney in Rosheim brought home to me that I was in a place the like of which I had never set foot in before.

The second incident was memorable for a very different reason. One evening, shortly before sunset, while my companions were deep in theological discussion, I set out to explore as much of the environs of the city as I could before darkness fell. I wandered into the low wooded hills on its western side. There I found the most extraordinary passages, retaining-walls and tunnels. Only when I came up against a massive gun did the thought cross my mind that I had probably unwittingly strayed into the Maginot Line. Not a soul was to be seen. The *Marie Celeste* was not more deserted. The Germans were on the very point of invading Czechoslovakia and the whole of Europe was alleged to be trigger-happy, yet here was a part of the great national defensive system of Eastern France apparently unmanned; and sure enough I made my escape back to the city without a challenge of any sort.

From Strasbourg we travelled by train down the Rhine Valley, stopping for a night at St Goarshausen and feeling unexpectedly and rather awkwardly safe in Hitler's Germany. But it was the safety of a bird of passage, glad to find a night's resting-place but quite ready to move on in the morning. The Rhine Gorge, I need hardly tell you, made an unforgettable impression. I had already encountered it through innumerable pictures reproduced in books and magazines, through the poetry of Heinrich Heine, which I had been required to commit to memory in the short introductory German course I had taken in the Sixth Form, through a newly acquired interest in the music of Wagner (which was not destined to develop much further) and in the study of geomorphology (which was), and through the brochures of the travel agents, which I had come to think of as addressed to other people, but not to me, not, that is to say, until now.

Our final destination was the Benedictine monastery at Amay-sur-Meuse in Belgium, which also had a special connection with the Russian Orthodox Church in exile. There, in the middle of the night, we shared in the Easter Services, very moving, very mysterious, very long and very suggestive of the idea that, whatever religion is about, it hasn't anything to do with this world and the everyday things that happen in it.

By the time we reached Ostend to take the boat to Dover I felt rather like a child who has for years been struggling to piece together a jigsaw puzzle of a few dozen pieces with reasonable success and has just been presented with one of several thousand of which the first few had just begun to fit into place. To have made a start was something, but the overall impression was one of a huge challenge ahead. A whole new world lay beyond the Straits of Dover and in it a seemingly infinite repository of raw material from which to build my own. How much of it, I wondered, could one assemble in a lifetime?

A more immediate question was when could I resume the task of exploring this new world beyond the sea. I didn't know then that it would be another twelve years before I would set foot on the Continent of Europe again, twelve very different years from the nineteen which had comprised my life up to that moment.

4   FACING THE REAL WORLD

*'Why with such earnest pains dost thou provoke
The years to bring the inevitable yoke . . .?'*

The King's Uniform

The declaration of hostilities between Britain and Germany in
September, 1939, found me wholly unprepared to cope with a
problem of far great magnitude than any I had had to face before. At
nineteen I was socially, politically and temperamentally immature for
my age. The basic ethics of Christianity had been well and truly
inculcated, even though the metaphysical aspects of its theology
were already being seriously questioned; and here was I, like
everyone else in my age-group, confronted with a challenge to
reconcile this whole ethical system with a pattern of behaviour which
seemed to run counter to its most basic principles.

The discovery that all morality is relative, that there is no
such thing as absolute good and absolute evil, came too late to
help me through this imminent dilemma. For the immediate
future the way was clear enough. During the early months of the
war I was pre-occupied with working for a War Degree, an
academic qualification devised by the University to enable its
junior members to put letters after their names before going off to
put naval, military or air force ranks in front of them. The hurdles
were lowered to a level at which a pass mark would be within
reach of anyone of reasonable ability, nevertheless we worked like
beavers, just in case.

Oxford, in those closing months of 1939, seemed suddenly to
have died. The blackout, which made it quite an adventure to visit
one's friends in another college, or even in one's own, after dark, had
a symbolic significance which spread much wider. Hitherto the
Oxford way of life, in which leisure and pleasure took their proper
places beside industry and academic ambition, had appeared to

require no justification. It was self-evidently worthwhile. The only thing wrong with it was that it would one day have to come to an end. Now things looked quite different. People began wondering what they were doing in Oxford at all. The place was pervaded by a kind of waiting-room atmosphere as though the one fact that had brought everybody together was that it wasn't quite time to do the next thing. But what was the next thing?

The religious context in which my moral code had been developed had ensured a heavy emphasis on the behaviour of the individual rather than the group. It was concerned with seeking the salvation of the soul, and to that end set up a series of rules and regulations which were manifestly incompatible with killing people. That they might also, in some circumstances, be incompatible with not killing people was too difficult a concept to fit into my slowly maturing philosophy. To contract out was no solution; to contract in would be intellectually dishonest, and therefore no solution either. Somehow I had to find a compromise which would allow me to get involved in something useful without simply throwing overboard the ethical misgivings about which I could not deceive myself, however effectively I might conceal them from other people. For most of my friends there was no problem at all in reconciling the positive action of enlisting with the preservation of self-respect. It amounted to the same thing. But for me the paradox was too challenging to be side-stepped in this way. Eventually I came to the conclusion that there was only one possible compromise. I registered as a conscientious objector and joined the army.

The Non-Combatant Corps was a very peculiar part of the military machine. We were not allowed to be allocated to what were called 'combatant duties' but otherwise we were available for a very wide range of tasks. During the first few months nobody seemed to know what to do with us and we were switched around from one job to another. The least interesting and least demanding (in skill if not in endurance) was cleaning out the latrines; the most physically exacting was carrying bags of sugar, 140 pounds I believe, for eight hours up sixteen steps to a storage loft. Other tasks required more skill. Thus I learnt the art of plate-laying while building railway sidings in Cambridgeshire and Worcestershire. I learnt to overcome any tendency I might have had towards vertigo and to walk along nine-inch girders four storeys above the street level while clearing up bomb damage in central London. I could well have been one of the first workmen to work on the Barbican site some decades before one of London's most prestigious buildings was designed. I even earned my living for a few weeks, I suppose one might say, as a professional comedian, being seconded with four others, one of whom *was* a professional comedian, to form a concert party, entertaining not only the troops of Eastern Command but, as a public relations exercise, other deserving communities and institutions. That's how I came to play an Ugly Sister to packed houses in various of the 'village colleges' in the vicinity of Cambridge and to find myself, quite literally, spending Christmas Day in the Workhouse.

During those early months, mostly spent in the vicinity of Cambridge, I made new friends whose company turned what could have been a period of unrelieved tedium and frustration into an enjoyable and rewarding one. My experiences at the Glebe House had equipped me better than most of my fellows for a restrictive life-style. I strongly disliked, as I always had, not being allowed to go where and when I chose, but previous familiarity with enforced confinement had taught me how to make the most of any opportunities which might arise for achieving at least partial liberty. We were allowed out in the evenings if we weren't on duty, as we usually were, and, while most people made for the nearest pub, I used to walk the lanes around Madingley, where the American Cemetery now stands, and follow them onto the higher ground. My favourite destination was a section of the Cambridge-Bedford road where it commanded wide views over the low-lying levels of the Fenland Edge. It was from that time that I can date the beginning of a changed attitude towards a part of England for which those early journeys to Bradley had built up, if not an antipathy, at least no positive liking. Seen from those low hills west of Cambridge, with the distant towers of Ely Cathedral rising out of their horizontality, the Fens suggested a new and expansive feeling of liberty, perceived if not attained. They had acquired the power to fascinate though perhaps not to comfort.

Early in 1941 there came an opportunity to do something more rewarding. It was to the great credit of the British Parliament that it made provision for us to contribute practically to the needs of the country rather than clap us all in gaol or shoot us, as some other governments might have done; but it was perhaps even more to the credit of the War Office that it took the imaginative step of inviting us to get involved in work which had hitherto been (quite unnecessarily) the exclusive preserve of combatant soldiers. Volunteers were called for to work with the Royal Engineers in bomb disposal squads, and here we found a job which was both useful and challenging and which allowed us to push any moral arguments, at least for the time being, somewhat into the background.

After a three-week period of training at Newark I found myself posted to No. 25 Bomb Disposal Company, Royal Engineers, which covered the south-eastern sector of London, obviously a busy one, because this was the direction from which most air raids were mounted. The dangers of explosion were much exaggerated in the public mind, though I confess this was not apparent when we first arrived at our quarters to find that one of the officers, The Earl of Suffolk, with some other ranks, had been blown up earlier that same day. We soon learnt that a much greater source of danger lay in the possibility of having a few tons of London Clay fall in on top of us, so the geology became a matter of more than academic interest. If at the outset we couldn't distinguish between the London Clay and the Blackheath or Oldhaven Beds we very quickly discovered the difference when the hole began to fill with water. I learnt the geology of the London Basin the hard way.

I also used this opportunity to learn something about the geography of London, this being the first time I had lived there for more than a few days at a time. People brought up in the city would probably have seen it quite differently, but for me the whole phenomenon of London took on the nature of an organism, a vast creature whose vital organs, situated some ten miles away to the north-west, were surrounded by a fringe of peripheral tissues connected with the centre through a system of nerves, veins and arteries. These were represented by a number of standard symbols of suburbia - Southern Electric trains, Green Line Buses, even trams in those days - and they connected up a succession of minor centres in which the distinctive architectural style of the nineteen-thirties paraded itself in numerous little shopping complexes usually around the railway stations. These were nodal points in a kind of cellular structure, each much the same yet each uniquely individualistic. The picture could hardly have been more different from the tree-animal on the Stibbard Turnpike, yet it evoked an equally animate image.

The animistic theme was even more potently and dramatically expressed in the job itself. The object of our quest very easily acquired a personality, but, like many villains, not a wholly unattractive one, and if it didn't exactly inspire affection it commanded respect. Trying to recover a big bomb from thirty feet below the street-level of London aroused some of the same emotions which I imagine must attend the hunting of big game, or maybe a giant fish, whose sleek, streamlined shape it more closely resembled. Certainly our quarry was static, but that didn't mean it was easy to catch. If it couldn't run or swim away it had frequently done its best to put us off the scent by pursuing a very devious path after entering the ground, and when we eventually caught up with it, if it lacked the menacing aspect of a 'Jaws' it packed into its belly a more devastating potential than a whole flotilla of Moby Dicks. Its pursuit aroused all the primitive instincts of man the hunter, and its eventual capture was attended by a sense of triumph such as might have followed the slaying of a dragon. If the relevance of this to the subject of this book escapes you, bear with me, because we shall be reverting to the question of animism in more detail later.

The chase itself was a progressive event moving through a series of stages and sometimes lasting for a matter of weeks. It began with the surface evidence. As it ploughed its way into the ground the bomb left behind a kind of tubular shaft and sometimes the top of this could still be seen, but nearly always the sides had slumped and filled it with mud, clay or other softer material. By pushing a steel probing rod down the filled-in shaft it wasn't difficult to establish the direction in which the bomb had gone, at least for the first few feet, and on the basis of such information as could be obtained the officer in charge made a decision where to sink a shaft which had to be timber lined as it was deepened (Figure 25). Quite commonly the bomb would have altered course as it reached greater depths, occasionally even beginning to rise again towards the surface. So we often had to build galleries, that is to say horizontal passages, outwards from the original vertical shaft. Not surprisingly it could be very dark down there, and working in a confined space by artificial light, with only room for one man to dig at a time, contributed to the sense of close confrontation in single combat with a hitherto unseen adversary. Incidentally, familiarity soon led me to overcome a mild tendency to claustrophobia at least to the extent that I could cope quite comfortably with these conditions.

The actual moment of discovery was always exciting. Invariably we had warning of the exact position of the prize before we could see it because it could eventually be reached by the probing metal rod, but the clearing away of the last veneer of mud and the exposure of the first tiny part of the curved carapace (Figure 26) was extremely rewarding. Like the members of an Everest expedition, wondering who will be selected to attempt the final assault on the summit, everybody hoped it would fall to his lot to be at the face when the moment of revelation arrived, so as to be the first to confirm the size of the creature. The next critical objective was to determine the type of fuse, because this had a bearing on how one set about the final stages of the operation. If, for instance, it was a time fuse, it was important to know. The fuse, or fuses - there might be more than one - were always in the side of the bomb, and it might, of course, be on the further side, so there could be much digging still to be done before it was eventually exposed.

The removal of the fuse was analogous to the kill. That was the tricky part. If there was going to be a bang that was the likely moment, and the *coup de grâce* was always administered by the officer-in-charge alone. Everybody else was sent back behind cover until Saint George emerged proudly displaying the dragon's heart. But although the beast was now, to all intents and purposes, dead, the recovery of the body could still be a hazardous process. These things often weighed 250, 500 or even 1,000 kilos (the last being

FIGURE 25. 'SOMEWHERE IN SOUTH-EAST LONDON, 1942'. Timbered shaft sunk to retrieve a 500 kg unexploded German bomb (see Figure 26). The object partly obscuring the view is one of the skips used to raise the excavated material. Photo by the author, 1942.

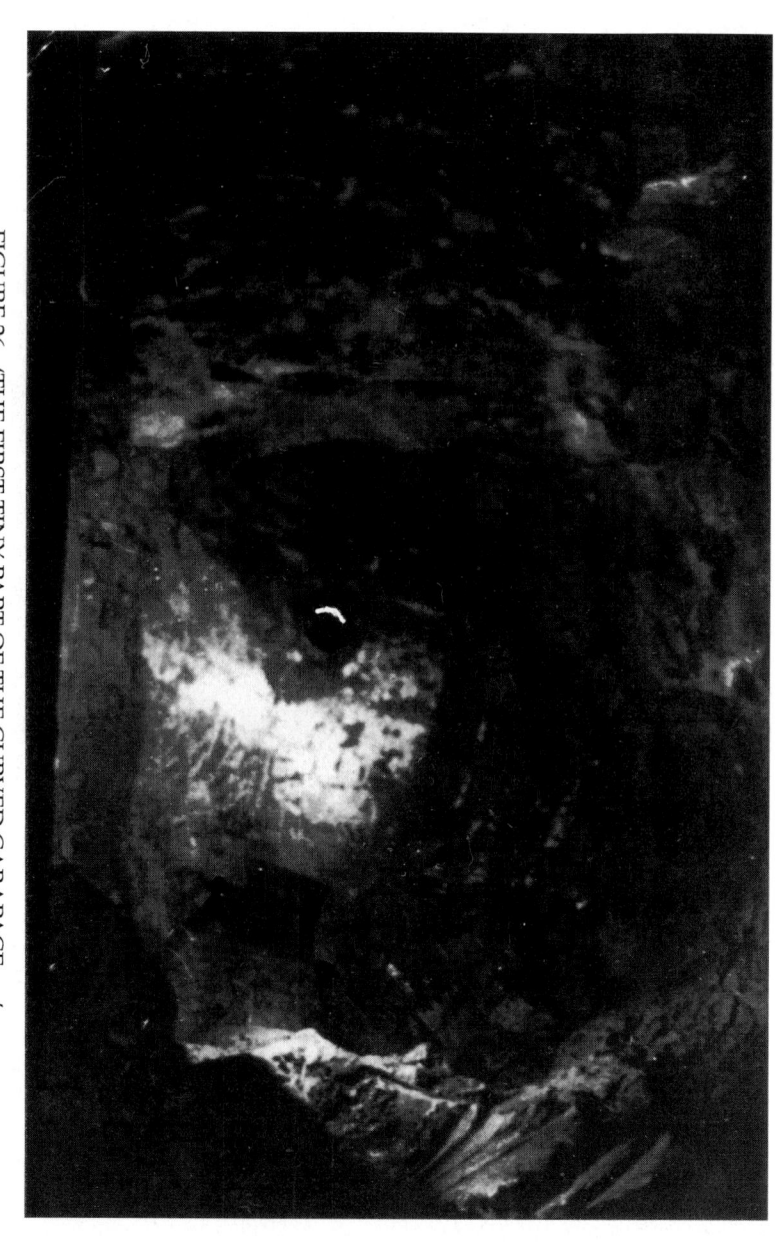

FIGURE 26. 'THE FIRST TINY PART OF THE CURVED CARAPACE . . .'
A 500 kg German bomb embedded in London Clay and partly exposed. The bright speck of light towards the left is the reflection from the steel locking-ring which holds the fuse in position. This particular bomb was reached by a horizontal gallery driven from the bottom of the shaft shown in Figure 25 at about thirty feet below ground level. Photo by the author, 1942.

equivalent to a ton) and occasionally even more, and the manipulation of a carcass of this size in the restricted confines of a hole in the ground, particularly if it involved movement along horizontal galleries, taxed the ingenuity of the sergeant or whoever was in charge of that part of the operation.

Eventually the beast was landed and lifted on to a utility truck or a 3-ton lorry to be taken away to the Thames Marshes, and sometimes it would fall to my lot to accompany it. There it would be placed in a shallow pit in the alluvium, plastered up with an explosive charge and connected by wires to a safe place of refuge some distance away where we took cover and dispatched it to Kingdom Come! The only hazard then was to brave the shaking fists of the housewives of Erith who, we were told, had been known to lose a window or two on such occasions.

During the years while it lasted, this job had a certain compulsive fascination. I can't say I was sorry when it came to an end, because that was the consequence of a spectacular drop in the number of air raids on London as the allies gradually achieved a supremacy over the *Luftwaffe*, but as a life-style it was preferable to the previous relative inactivity, and it was a period in my life which I can look back on with a certain nostalgic satisfaction. I wouldn't have missed it, and I met some of my best friends there.

As the number of bombs dropped on London diminished so the number of prisoners of war captured in foreign fields increased and many of them were brought back to Britain and I spent the later part of the war working in a prisoner-of-war camp. I foolishly failed to take full advantage of the opportunity to learn more than a smattering of German and Italian, largely, I think, because Edmond Erranti, the prisoner detailed to be my clerk, was a French-speaking Swiss who had been unwise enough to visit his Italian grandfather and had consequently found himself on an Italian parade-ground unable to speak a word of the language in which commands were given. He also spoke little English, and because I was already able to converse at a modest level in French it was natural that this was the language we used in our day-to-day work. Although I never became fluent, I did reach a stage where I could conduct a reasonable conversation. Edmond became a good friend and I have subsequently on more than one occasion taken my family to stay with him on Lake Geneva. I suppose it can hardly be usual for a guard to spend his holidays with one of his former prisoners.

Just as at school I was periodically allowed to return to that private family life which went on all the time, whether I was there or not, so in the army I was kept in contact with home by my statutory entitlement to apply for leave. There was no entitlement actually to *have* leave, only to apply for it, but I was generally lucky. I was constantly reminded that others were not so fortunate. In particular my sister's fiancé, Bill Cook, spent years in North Africa and Italy with never a sight of his bride-to-be. They solved their problem of communication by writing letters at every possible opportunity, no less than six thousand in four and a half years! All of them survived, and forty years later they published their story, using the letters as their primary source. Bill was an army chaplain, hence the title *Khaki Parish*, (Helen and Bill Cook, 1988) and on the strength of those six thousand letters they found themselves in the *Guinness Book of Records*.

During those six years and six days when I wore the King's uniform several events took place which enormously affected my life, the first of them, to be precise, just before I joined up. My father was offered an appointment in a parish roughly ten times as large as Stibbard, and he was then so far recovered from his malaria that he had no hesitation in taking it on. So in 1940 he became Rector of Diss, some 436 years after the Poet Laureate, John Skelton, had assumed the same office, and Stibbard ceased to be 'home'.

The old market town of Diss was also in the County of Norfolk, but only just. Its southern boundary is formed by the River Waveney which is also the County boundary; on the further side lies Suffolk, the county of Gainsborough and Constable. I found Diss an attractive little town with its romantic mere and its picturesque streets. The Rectory was a half-timbered house of considerable charm and its garden as full of potential for the stimulation of the imagination as that of Stibbard. It had a tennis court, a large greenhouse with a productive vine, an old mellow wall, a pond, a mulberry tree and a splendid group of tall beeches as well as a conspicuous Wellingtonia. But the criteria by which a twenty-year-old judges his surroundings are not the same as those which commend themselves to a small child. Powerful as were the responses which the garden at Diss was capable of evoking, they were responses which had already been developed elsewhere. Its formative influence could in no way compare with that of Stibbard which I was not to see again for some years.

The next relationship to be severed was not with a place but with a person and one to whom I had been very close. We already knew that my mother was ill when we moved to Diss. There she was able to enjoy the garden and to make new friends among my father's new parishioners, but her strength was failing and she died about a year later. By that time many other influences had been at work helping to shape my tastes and preferences, but none of them approached in importance the influence of her kindly guiding hand. She never expected me to replicate her own ideas, and there were many issues on which we took somewhat different views, but they seemed to be differences of modes of expression of the same basic attitudes. Her influence on me was not to cease, but there were to be no fresh infusions of her ideas, only the continuation of an individual life which she had so directed through its most formative years that nothing it subsequently accomplished could be said to be wholly unaffected by her. It says something of the extent to which attitudes have changed that nobody ever told me she died of cancer and it was many years before I found out.

In addition to the ordinary periods of seven-day leave and occasional weekend passes I was successful in the summer of 1942 in obtaining a longer period, twenty-eight days in fact, to help in the harvest under a scheme agreed between the War Office and the Ministry of Agriculture. The arrangement was negotiated through the East Suffolk War Agricultural Committee (the WARAG), and as a result I found myself working for a month at Wortham Manor Farm, just on the Suffolk side of the Waveney and about five miles from Diss Rectory where I was able to live.

This was the same farm which, a few years earlier during the great agricultural depression of the 'Thirties, had achieved national fame as the headquarters of the 'Tithe War'. The farmer, R. H. Rash, had been a leading campaigner for the abolition of tithes, the system of payment by which the Church of England had been chiefly financed from medieval times onwards and which, though well on the way to being superseded, was not yet dead and buried. It was claimed that this had been the last straw in bankrupting many East Anglian farmers, some of whom I had known personally. Although the Tithe War represented a fundamental challenge to the institution in which my father earned his living, he had established a friendly relationship with Rowley Rash for whom he had a high regard which I believe was reciprocated, and this, no doubt was how Rowley had come to ask for me personally to be given leave of absence to help him out. Another circumstance which clearly helped was that my temporary employer was a very influential member of the WARAG!

Indeed when I was working on the farm it seemed that Rowley himself spent almost every day somewhere else in the county on WARAG business and I generally found myself lunching alone with Mrs Rash. She was already well known as the novelist, Doreen Wallace. I was keenly aware from the outset that she was my intellectual superior, but she was extremely kind to me and I greatly enjoyed the stimulus of those lunchtime conversations. Years later I learnt a lot from her book, *East Anglia*, about the art of topographical description.

The job itself provided me, after years of living in the country, with my first experience of actually working on the land. Harvesting in those days was dusty, dirty and very exhausting, but in a way immensely satisfying. The process was affected by the weather, the acreage to be cut, the yield, the type of crop, the size of the available labour-force, and so on, and my own job varied accordingly. Combine harvesters being still a thing of the future, reaping was carried out with a mechanical binder which cut the stalks and bound them in sheaves. This was a skilled task, not one for me. I was usually given the job of carting the sheaves to the stack or, if it was available, straight to the threshing-machine. This was always called 'leading', because we still used a large cart-horse, Prince, the only horse I ever came to know personally. The threshing process involved feeding the sheaves into the top of the drum (see Figure 6), bagging the grain and building a straw stack from the straw which was delivered up the elevator. The bagged grain had then to be carted away to the barn. We still called this 'leading', though we used a tractor to do the job. From all this it will be apparent that a fairly large labour force was required, which is why I managed to get four weeks leave from the bomb disposal squad.

This experience brought home to me that the separation of urban dwellers from the process of producing the food they eat deprives them of one of the most elementary and basic sources of human satisfaction. To play even a minor role in the corporate act of going into the fields, collecting the bounties of nature and claiming them for the sustenance of one's fellow creatures is to encounter one of the most meaningful expressions of the relationship between people and their environment. The lesson lasted no more than a month, but it was not to be forgotten.

During that period of 'harvest leave' my sister, Helen, was also living at Diss Rectory, writing prolifically to Bill Cook, though just how prolifically I was not to discover for forty years! For the previous three years she had been a student at the Froebel Educational Institute which was normally situated at Roehampton in South-west London, but had been evacuated for the duration of the war to Knebworth House in Hertfordshire, home of the Earls of Lytton and much later destined to become the venue of one of England's largest pop festivals. During these holidays, and before starting her first teaching job, she invited a college friend, Iris Hearn, to stay for a few days at Diss. At the conclusion of her visit Iris missed the train home and had to stay another night. In the event she stayed another fifty years and is happily still with me.

Iris and I were married a year later at the church in Hatfield Broad Oak, Essex, where her father had been Vicar, so yet another connection was forged with the Church of England. He had died shortly before, so to my regret, I never met him, neither did Iris meet my mother. *Her* mother had died many years earlier when Iris was five.

The early years of marriage were inevitably years of intermittent separation. Iris was teaching at the West Norfolk High School for Girls at King's Lynn (though she taught little boys as well in the younger age-groups), and after two years she moved to the Abbey School, Reading, where she had been a pupil. After leaving the bomb disposal squad I spent a short time at a P.O.W. Camp at Ledbury in Herefordshire. This was a cosy little country of hills and woods, and while I was there Iris would join me as often as possible. We walked extensively in the area and particularly on the Malvern Hills to which we became deeply attached. Before I was demobilised I secured a transfer to a P.O.W. Camp at Botesdale, quite near Rash's farm at Wortham Manor and near enough for us to be able to set up our first home in Diss Rectory. Iris' brother, Bob Hearn, also moved his base there when he left the Royal Scots Greys and went to Trinity College, Cambridge, and with my sister, Helen, still based there until Bill Cook's return, my father's house remained very much the centre of the family as we turned to face whatever the condition of 'Peacetime' might bring.

I was eventually demobilised in June, 1946. Those six years as a private soldier had been an important part of my educational experience and left me in many ways a different person to face the next challenge, namely what to do with the rest of my life. Decision-making this time was going to be constrained by a number of issues which had not been relevant before, and not least by the fact that I now had the responsibilities of a family man, young Richard having been born the previous year.

Civvy Street

Like many an ex-serviceman about to be demobbed I had a better idea of what I didn't want to do than of what I did, and eventually I found my thoughts increasingly turning in the direction of an open-air life. My brief taste of farming at Wortham Manor had been a pleasant one, and after a long discussion with Rowley Rash I eventually decided to take up a way of life which, though more specialised, would incorporate many of its attractions. Under a training scheme operated by East Suffolk County Council I became apprenticed as a trainee to Arthur Holmes, Quaker, fruitgrower and thoroughly nice person, and under his expert and friendly tutelage began to prepare for whatever the future might bring.

My purpose in going to Arthur Holmes was to learn the art of fruitgrowing rather than other kinds of husbandry, nevertheless I did have the opportunity to learn a little about the more general aspects of farming. Holly Tree Farm, Bramfield, near Halesworth, had been a small mixed farm typical of the agriculture of that part of East

Suffolk. When Arthur Holmes had purchased it he had immediately started to plant it out with fruit, mainly apples but some pears and plums and a smaller acreage of soft fruit, chiefly blackcurrants. The conversion, however, was not yet complete, and there remained fragments of residual enterprises belonging to its former function.

So my first job in the morning was to milk two cows and to feed the pigs. Occasionally I had a chance to tackle some other jobs on the farm. Once or twice I was even allowed to plough behind two horses, but this was a job jealously guarded by the senior farm worker who had been doing it for decades. The particular type of ploughing was unusual for the twentieth century. The so-called 'ridge and furrow' had been the basic system of ploughing in the open-field arable agriculture of much of North-western Europe, but had been virtually eliminated in England by the Inclosure Movement of the eighteenth and early nineteenth centuries. It was an inefficient method because, although quite good cereal crops could be grown on the well-drained ridges, the water which stood in the troughs or furrows effectively drowned any seed which fell in them. The system survived at Holly Tree Farm only in one heavy wet field where bad drainage precluded the use of any other method.

When other tasks were not urgent I acquired some experience of hedging and ditching and doing the hundred-and-one things that have to be done on any farm, but most of my time was spent among the fruit trees and bushes. As with other agricultural activities the work pattern was seasonally varied - pruning, spraying and planting in winter; spraying again, cultivating and muck-spreading on blackcurrants in the spring; picking in the summer and early autumn and fitting innumerable little jobs in between. Some of the work had to be carried out communally; there were four or five regular employees and occasional extras were taken on at busy times, but often I would spend a whole day working in an isolated part of the orchard and hardly seeing anyone else, so I had some more practice in learning how to live with my own company.

I particularly recollect those dark winter days of pruning when one could hardly see as far as the limits of the misty orchard. There was a certain awareness that some murky visual contact could be established with the outside world but only along straight, fixed channels between forests of twigs, which, even though bare and leafless, were massed thickly enough along the lines of the trees to blot out every object below the sky and even the lower part of that. It was a sort of chessman's-eye-view of the world (Figure 27), controlled as rigidly as a bishop's move by the lines of the planting.

I remember thinking of a passage from Virgil's *Georgics*. This is how, in Dryden's translation, he describes the way to set out a vineyard:-

FIGURE 27. 'A SORT OF CHESSMAN'S-EYE VIEW OF THE WORLD'. The gable of Holly Tree Farm, Bramfield, can be seen along one of the rows between the dessert apple trees. View to the south. Photo by the author, 1986.

If fertile Fields or Valleys be thy Choice,
Plant thick, for bounteous Bacchus will rejoice
In close Plantations there; but if the Vine
On rising Ground be placed, or Hills supine,
Extend thy loose Battalions largely wide,
Opening thy Ranks and Files on either side:
But marshall'd all in Order as they stand,
And let no Soldier straggle from his Band . . .

So let thy Vines at Intervals be set,
But not their rural Discipline forget:
Indulge their Width, and add a roomy Space,
That their extremest Lines may scarce embrace:
Nor this alone t'indulge a vain Delight,
And make a pleasing Prospect of the Sight:
But, for the Ground itself this only Way,
Can equal Vigour to the Plants convey;
Which crowded want the Room, their Branches to display.

Whether Arthur Holmes knew his Virgil I can't say, but this is how we did things down at Holly Tree Farm a couple of thousand years later.

There was a certain solitary satisfaction in just getting on with the job of winter pruning, closeted from the rest of the world. Planting tended to be a more sociable occasion because it could be done most efficiently if the planters worked in pairs, one to hold the tree while the other filled the hole and tamped down the loose soil; but the most ritualistic social occasions were reserved for the picking season. Generally a group of two or three pickers would move from tree to tree bringing the fruit to the farm trailer to be carried off to the barn.

The atmosphere in the barn at this time I can only describe as 'heady'. Quite a small bowl full of Cox's Orange Pippins gives off a pretty penetrating aroma. Imagine hundreds of boxes full of them stacked in layers, literally tons of aroma-exuding apples! The very air was intoxicating, and two or three deep breaths made the nostrils prickle with physical pleasure. There was, too, the same feeling of collective achievement which I had experienced 'leading' the cereal harvest at Wortham Manor or retrieving an outsize German bomb.

The work at Bramfield also gave me my first experience of actually changing the landscape on something more than a minuscule scale. The orchard in ten years time would carry the imprint of my own hand, because it's by pruning that one controls the shape of the growing tree. The model we were encouraged to bear in mind was that of a wine-glass with a short stem from the top of which the curving sides spread outwards and upwards to an even level rim. To look back along the row at the end of a day's work and see a straight line of arboreal wine-glasses tailing off into the distance gave one a certain sense of involvement in an order-inducing process. That the resultant pattern of extreme regularity was the very antithesis of what I more usually enjoyed in landscape was to provide just one more challenge when I later attempted to make rational sense of my own preferences, never mind anybody else's.

So what did I more usually enjoy? I shall be making reference later to my favourite types of landscape, but at that very moment I was living right in the middle of one of them. After the death of my mother in 1941 my father had begun to give some thought to his own future. As a clergyman he had always lived in a house which went with the job, and, while this arrangement had some advantages, it meant that, when he eventually retired, he would have to leave Diss Rectory vacant for the next incumbent. He therefore decided to acquire a house with the ultimate intention of retiring there, and in the meantime it would be available for use by any members of the family or by any friends who might like to make use of it for holidays or any other purpose.

At that time, late in 1941 or early in 1942, house property within a mile or two of the East Coast was being sold very cheaply because it was widely believed that it would be reduced to rubble as soon as the invasion came. Consequently my father was able to pick up a remarkable bargain. The name of the house was Fen Cottage, and, since it was only thirteen miles from Bramfield, for two years from the late summer of 1946 it became home for Iris and me and little Richard, who spent those highly formative years from one to three in this rather striking environment. I must tell you more about it.

The house itself (Figure 28) was a detached building of some elegance, roofed in the traditional pantiles of the area and finished in a cream-washed rendering. It had been designed just before the First World War by a local architect, Cecil Lay, whose romanticism was not always matched by his sense of the practical. It had originally been flat-roofed, but the pitched roof had been added long before we saw it. Its three bedrooms were reached by an impressive staircase in an entrance hall which must have comprised a fair proportion of the total enclosed space of the house and which, being on the seaward side of the building, had a wonderful capacity for losing heat in winter. The whole of the western side on the ground floor was occupied by a long, low lounge with windows too small for the size of the room. Most of the time it looked dark and rather gloomy, but when transfixed by shafts of the evening sunlight it became suddenly alive with dancing leaf-shadows and extremely beautiful.

FIGURE 28. FEN COTTAGE, ALDRINGHAM COMMON, SUFFOLK. Note the hedge of *Cupressus macrocarpa* (left), now even taller than in the nineteen forties, largely obscuring the view, the association of Scots pine and silver birch (right), and the bracken in the foreground, separated from the garden by the corner of a ploughed field in light, sandy soils. View to the south. Photo by the author, 1986.

FIGURE 29. DUNWICH HEATH, SUFFOLK. The open landscape of heather and bracken with sporadic shrubs typical of the light soils of the Suffolk 'Sandlings'. View to the north-west. Photo by the author, 1986.

The house was surrounded by a triangular garden. It contained a row of my favourite pines and it was enclosed on the north and the south-east by a hedge of *Cupressus macrocarpa* some twelve to fifteen feet high and totally opaque. You'll understand, therefore, how easily the castle imagery was re-awakened. The garden sheltered by the hedge was extremely cosy, visually and microclimatically protected from the outside world.

But it was this same outside world which made the place so important for me aesthetically and emotionally. There are in East Anglia many places where heathland associations of silver birch, heather, gorse, bracken and the occasional Scots pine thrive on light, sandy soils. I had already become addicted to this kind of country at Pretty Corner. One patch of it stretches in a wedge for some miles along the Suffolk Coast (Figure 29). It had furnished the rough, uncultivated landscape for Crabbe's *A Lover's Journey* (Crabbe, 1812) in the nineteenth century and was already the spiritual (and indeed the actual) home of the composer, Benjamin Britten, during those years when I was working at Bramfield.

Fen Cottage stands, albeit now under a different name, on a low plateau of this type of heathland which rises gently out of the coastal marshes some two miles north of Aldeburgh and a mile inland from Thorpeness. Although the view from the ground floor was totally blocked by the cypress hedge, the upstairs windows commanded a wide prospect across the marshes to the sea. It was the juxtaposition of the house within its enclosed garden and the open landscape beyond it that made possible the feeling of being at one and the same time within a protected area yet able to survey the surrounding countryside. The little plateau itself, on the landward side, was just fertile enough to support poor crops of rye, but on the seaward side it carried a wilderness of gorse and brambles (in September a veritable cornucopia of blackberries), crossed by a few sandy tracks which served as our lifeline across Aldringham Common to the public road and the rest of the world.

When I think of it now, as I often do, the picture which comes to mind is of a summer evening a little after sunset. The pines cut a black *silhouette* against the western sky. The marshes in the east are drained of light, and one can discern the vague outlines of a few clumps of willow and poplar which rise out of the unseen levels to hide the open water of Thorpeness Meare, all that has been artificially preserved out of the medieval North Harbour of Aldeburgh. The bird-calls of the daylight hours - the little song-birds of the heathland the corncrakes in the ryefields - die away, the last to go being the booming of the bittern in the reed beds, and the heath is left to the song, if that's the right word, of the nightjar. With luck it may be possible to discern one in the fading light among the gorse bushes churring away like a sewing-machine.

One of the reasons why the sight and sound of these birds were particularly memorable was that I had only recently begun to learn a smattering of ornithology from Iris, who, like a good teacher, successfully communicated her enthusiasm as well as her knowledge to an eager pupil, and a new dimension was added to my understanding of the common which could be simultaneously savoured with all the senses. It was she also who first gave me a real appreciation of the significance of the idea of 'habitat' which assumed a great importance in my thinking later on. I had come across the idea when I had been to Hornby as quite a young child and came across birds like curlew which I had never seen at Stibbard, so the idea that particular species naturally frequented particular kinds of landscape was not new. Later I had met the otter on the tree-trunk and applied the same thinking to a very different kind of environment; but at Fen Cottage I was able to acquire a much more detailed, a much better-informed understanding of this phenomenon.

If all had gone according to plan my two years of training at Bramfield should have led to my acquiring a small holding and planting it out with fruit trees. Had I got as far as committing myself to such a course I should almost certainly have run into problems because the market was becoming increasingly difficult as the import of foreign-grown fruit was resumed. It wasn't the market, however, which eventually forced a decision to change course. In the spring of 1948 I experienced pain and swelling in the ankles. Within a week the condition had spread to both knees and soon afterwards to the hips. I spent some weeks unable to do more than hobble around on a couple of walking sticks. I discovered that arthritis is not uncommon in people under thirty, and, as the weeks dragged on into months, I began to have second thoughts about the wisdom of pursuing a career as a fruit-grower. Although the condition was eventually cured the doctor couldn't promise it wouldn't return; so I made another turnabout. I applied for entry to King's College, Newcastle-upon-Tyne, which is now the University of Newcastle but was then a constituent part of the older University of Durham, and during the summer of 1948 I was accepted to read for an Honours Degree in Geology, taking the special two-year intensive course which had been introduced for ex-servicemen returning to embark on their chosen careers.

When I entered King's College I still had only the haziest idea of what I intended to do eventually. I must have had some notion of becoming a professional geologist, but again Providence intervened. I arrived in Newcastle in October, 1948, to find that the two-year Honours course in Geology for ex-servicemen had been discontinued and that I had been accepted by mistake! The professor who had accepted me had retired during the long vacation, so I had to negotiate with his successor, Professor Westoll, who could reasonably plead 'not guilty'. Although he went out of his way to be helpful, there were limits to what he could do. I was offered the chance to read the first two years of a three-year course simultaneously as long as I also made up the gaps in my basic knowledge of chemistry, physics and biology in my own time. In view of my essentially arts background at school, this looked a pretty tall order, particularly since time-table clashes between the first- and second-year programmes would make it impossible to attend all the lectures I was supposed to.

I got the message and went along to see Henry Daysh, Professor of Geography, explaining that I already had an unclassified War Degree at Oxford and wanted an Honours Degree, having begun to revive an earlier interest in the career of school-teaching. Thus it was that, ten years after I had enrolled at Oxford for a course leading to the Degree of Bachelor of Arts, I found myself once again an undergraduate embarking on another Bachelor's Degree, of Science this time, at another university, thoroughly redbrick, grossly overpopulated in the post-war conditions, but full of an exciting mix of school-leavers and ex-service men and women together. One of my fellow undergraduates from Oxford days, John House, was appointed to be my tutor and I embarked on the next stage of my rather unorthodox academic career. John handled what could have been a difficult relationship with skill and understanding. He became a good personal friend until he died in 1984 shortly before he was due to retire as Professor of Geography at Oxford.

As for geology, I was able to take it as my Subsidiary Subject, and at that less exacting level I enjoyed it immensely. I discovered in fossil-hunting the same sort of satisfaction that I had previously experienced in bomb-hunting, the main difference being that success came much more quickly and with much less effort and discomfort. A particularly pleasant obligation which it imposed on me was attendance at a week's field course that took me back to the hills of South Shropshire where I had found so much enjoyment a dozen years earlier. It was based on Church Stretton in the very heart of the Housman country, and the acquisition of a new technological language had to share my attention with the revival of old memories. One evening, when I should have been reading up the stratigraphy of the Ordovician in preparation for the next day's work, I put pen to paper in the Stokesay Castle Hotel. Apart from a couple of passages where I have tidied up the scansion, I give you the result as it appeared on the fly-leaf of my field notebook:-

Plough straight, my lusty team, and deep;
Fear not to wake the lads that sleep
With frigid maidens, deaf and dumb,
In Onny's broad alluvium.

The lad that loved when spring began
Sleeps in the Middle Cambrian;
A fathom deep he takes his ease
And lies with *Paradoxides*.

Oh, I have been to Ludlow town
And hack'd the Whitcliff up and down,
And quarried from the sullen stone
The steadfast and enduring bone.

And does my girl lie down and cry?
Nay, lad, her rosy cheeks are dry,
And other lads beside her sit
In April on the Hoar Edge Grit.

The ploughman who, on Corndon Hill,
Espies the blackbird's yellow bill,
Will find no stone to curb his flight
As hard as Corndon Dolerite.

Though Shropshire yeomen rise betimes
To bear the flag to foreign climes,
I'll rest content, when battles rage,
With flags of Upper Ludlow age.

Now soon the clock will strike for me
To hang in steepled Shrewsbury
And rest my thews for evermore
In Severn's wide meander-core.

The principal reason why I had chosen Newcastle as a venue for my resumed undergraduate career was that my Uncle Ted and Aunt Annie, the same whose neighbour's house I had misguidedly entered in Gosforth, had by this time moved south of the Tyne and were living in Low Fell where the upper storey of their house was not in use. Enquiries in Oxford had very soon established the difficulty of finding a place for a wife and child, otherwise the logical step would have been to go back to Christ Church. I therefore took up the kind offer of my aunt and uncle and reverted to the role of an attic-dweller which had been my lot on first entering the world at Headingley.

Throughout that first year at Newcastle my Uncle Ted was clearly ill and in the summer of 1949 he died. We succeeded in finding a former miner's cottage near Rowlands Gill in the North-west Durham Coalfield. It was built on the side of a hill and had three storeys at the back though only two at the front. Apart from a little lawn terraced out of the hillside and a few beds for growing flowers and vegetables, the garden was more or less a grass field with a spring in it and a small stream issuing therefrom and plenty of trees encroaching into it from the adjacent wood. It had therefore high potential for the development of young Richard's exploratory prowess. In spite of the still active 'bee-hive' coke-ovens, the last, I believe, to operate in England, at the end of the cinder-track which served as our only access-road, the general impression was of a rural landscape trying with some success to fend off the encroachment of a 'paleotechnic' type of industry.

I think it was during this year at Rowlands Gill that I underwent a radical change in my sense of place in that I first became keenly aware of what it meant to have cut the umbilical cord with East Anglia. I had lived away from home for a number of years, but I had always had the parental home to return to. I regarded the terms at school and at Oxford as safaris, excursions of two or three months duration away from the home base. Even in the army I was able to get home at intervals and latterly even to live there. My most treasured possessions were still there whether I was or not, and going there was always like going home, at least until I came out of the forces. My mother's death had left a big gap, but it had not destroyed my view of the parental home as my home. The base at Diss Rectory, where Iris and I were living when our first son was born, and which for that reason also has always held a special place in my affections, was still therefore to be visited more or less whenever we wanted to go there, but it was no longer 'home'.

Before I left for Newcastle my father had married again, but I very much doubt whether this had anything to do with my change of attitude towards Diss Rectory. We all became very fond of my newly-acquired step-mother, Elsie, but I now had my own family and had embarked on the process of establishing it in an independent place. For the two years when we had been at Fen Cottage this had still been in East Anglia, only an hour's drive from Diss, but the move to Newcastle, 250 miles away, powerfully reinforced the feeling of independence, and from now on home was to be wherever my family was and East Anglia began to feel like somewhere else.

Much as I liked Newcastle as a city, the people I met there and the little cottage at Rowlands Gill, and although I had no difficulty in thinking of it as home, somehow I could never quite conjure up the feeling that this was a place which I was going to be able to grow

into, as a transplanted fruit tree puts down its roots into the earth to
claim it as its own patch and become almost an integral part of it. It
could never establish a relationship like that which had bound me so
intimately to Stibbard. I suppose I knew that in any case another
move was likely to take place very shortly.

The coalfield which lay all around us was a curious mixture of
not-very-attractive mining villages, rather exposed farmlands on the
plateau and pretty, wooded, steep-sided ravines. To the west lay the
higher moorlands of the North Pennines which powerfully attracted
me, but we couldn't afford a car, and to reach them by public
transport with a small child was too formidable a task to be
attempted often, so for the most part we were penned into our little
oasis, an intimate little nest, almost surrounded by woods, in what
felt like an alien land; not unfriendly, certainly not hostile, but still
alien.

During my final year at Newcastle I had been encouraged by
Professor Daysh to believe that an appointment to a university
teaching post might not be beyond my reach, provided I could add a
measure of luck to the fairly hard work I had put in for two years. So
during the spring of 1950 I applied for an Assistant Lectureship in
Geography at the University College of Hull (Figure 30). The measure
of luck duly came my way and, although I still had no Honours
Degree to put after my name, never mind any research experience, I
was offered the job subject to my final examinations being satisfactory.
In those, too, my luck held, and in the summer of 1950 my full-time
education, interrupted as it had been by various vicissitudes, came to
an end just a quarter of a century after it had begun. I was thirty and
more than ready to get stuck in, at long last, to my chosen career.

My story so far has taken us less than half way to the present,
but I shall not need to burden you much longer with this
chronological account of the biographical details. The perceiving
organism, which is me, had by now matured in its essential outlines.
What lay ahead was rather in the nature of minor modifications - fine
tuning, if you like - which needn't hold us up now. So let's quickly
conclude the *curriculum vitae* in the barest outline. We shall then be
in a position to look more closely at this crude instrument of
environmental perception, with its preferences and its prejudices, its
fancies and its fallacies, its mixture of potentialities and limitations,
all legacies of a biographical record which is ordinary enough, but,
like everybody else's, yours included, unique.

In September, 1950, we moved into a small semi in Golf Links
Road, on the edge of the city of Hull (Figure 31). The golf links had
long ago ceased to exist. Its greens and fairways shared with a less
romantic rubbish dump the privilege of having furnished the site for
our house and those around it. We stayed there for nearly 10 years

FIGURE 30. THE UNIVERSITY OF HULL IN THE MIDDLE 'FIFTIES. This
photograph shows the campus shortly after the University College of Hull had
become an independent university. By this time the only major addition since my
appointment in 1950 was the large, light-coloured Chemistry Building (centre). My
office was in the nearer of the two huts partly obscured by trees on the extreme left
of the site. View to the north. Photo by Hull University Air Squadron, courtesy of
The Librarian, the Brynmor Jones Library.

FIGURE 31. GOLF LINKS ROAD, HULL. An inter-war suburban development on the outskirts of the city. Photo by the author.

FIGURE 32. 39 HULL ROAD, COTTINGHAM.
This Edwardian house has been home since 1960. Photo by the author, 1987.

before moving into a larger house barely half a mile away on the edge of Cottingham, a medieval village which liked to call itself the largest village in England but in fact was well on the way to being a dormitory for Hull. This house was still a semi, but of Edwardian vintage (Figure 32) with almost as many rooms as Stibbard Rectory and a half-walled garden large enough to provide some scope for the environmental education of our two younger sons. Charles had been born at Golf Links Road in 1957; Mark entered the world in 1960 shortly after we had moved to Cottingham.

Once or twice I toyed with the idea of making a move from the University of Hull (as the University College had become in 1954), but always in a half-hearted way. I was often warned of the dangers of allowing myself to get in a rut, but the fact is I found the university a very pleasant place to work in. My colleagues had become my friends - I couldn't ask for better - and there are always advantages in working within a system which one has come to know. We tended, therefore, to satisfy our *Wanderlust* by taking short-term appointments on secondment or special leave of absence at institutions in other parts of the world, but I remained on the staff at Hull until my retirement in 1982 and for another three years after that on a part-time basis.

It might seem logical at this stage to continue with this roughly chronological account, but there are procedural reasons why it will be better to switch to a different approach, making some references later to some of the events which happened after 1950 in the contexts where they appear relevant. The chronological sequence, the historical order in which the events of our lives take place, is the only sequence in which we can encounter them; but if we want to study more deeply their importance as formative agents in the processes by which the perceiver becomes the creator, taking his material from his experiences of the real world and building out of them his own unique model of reality, we shall need to re-assemble them, adding at the same time such other ingredients as may become necessary, in some meaningful order more appropriate to that task. We must look at concepts, topics and themes, tracing where we can their origins in our encounters with what we presume to call reality.

One thing, however, I must make very plain. If I have attempted to describe in Part I the processes by which, as its title implies, the world made me what I am, and if in Part II I am about to concentrate on those processes by which I have made a world of my own creation, this is a descriptive sequence imposed solely by convenience. The twin processes are essentially and perpetually interactive. The interaction started in infancy and has been going on ever since.

PART II I MAKE THE WORLD

5   THE WORLD-MAKING PROCESS

'See at his feet some little plan or chart,
Some fragment of his dream of human life,
Shaped by himself with newly-learned art.'

Making and Re-making

If you're interested in the philosophical problems of relating our own versions or models of the world to reality, I suggest you read *Ways of Worldmaking* by Nelson Goodman. 'If there is but one world', he writes, 'it embraces a multiplicity of contrasting aspects; if there are many worlds, the collection of them is all one' (Goodman, 1978, 2). I'm not going to follow him here into these deeper philosophical waters. Instead I shall take my cue from what he writes a little later (page 6): 'Worldmaking as we know it always starts from worlds already on hand; the making is a remaking.'

For my purpose the world already on hand is what I take to be the world 'out there'. It seems to exist independently of my observation or interpretation of it, but the problem is that everything I can know about the world has, before I can know it, already passed through the process of perception, my perception, and in doing so has already become 'remade'. Many of the episodes which I've described in Part I bear the marks of having been processed, not just by anybody but specifically by me; so our task now will be to look at some of the influences which have been at work during the processing. The opportunities for environmental observation which happen to have come my way may have determined what I have seen, but not necessarily how I have seen it.

All of us from time to time find opportunities for applying the creative process to the real world, actually changing its character and making its visual appearance different from what it would otherwise be; but for most of us the extent to which we can do this is strictly limited. We may plant a tree, remove a garden fence or paint the windows with a colour of our own choice, but even at this scale opportunities for changing the look of the landscape don't come everybody's way.

Certainly for members of some professions more ambitious possibilities arise. Architects, landscape architects and civil engineers may find themselves designing nothing less than entire environments, replacing landscape features, which have passed down from former generations, with new creations of their own devising. The Dutch water engineers have virtually created half a country, making new landscapes out of the marshes, the mudflats and the sea. Legislators have, in a different way, exercised creative roles, setting up the conditions in which planners can, at their several levels, prepare their blue-prints for contractors in turn to translate into the reality of landscape. The rest of us have to find some more modest ways of satisfying the creative urge.

One kind of creative activity which can hardly be described as modest, is that by which we reproduce our own species. Making the inhabitants of the future world can be one of the most far-reaching ways of influencing the development of what is to be their environment, and the satisfaction we derive from this is increased by the obligation it imposes and the opportunity it affords for participating in the shaping of the character of the organisms we have brought into existence. Procreation can be a long-continued process, at every stage of which we can experience the pleasures as well as the frustrations of the creative artist, and if I don't pursue further this aspect of creativity it is only because it leads us somewhat away from the environmental theme which is the subject of this book.

One aspect which is more immediately relevant to our enquiry is the making of small-scale models of the real world or more commonly some aspect or component of it. For a very few this may actually become a profession, but for many more it's just an important source of leisure interest. The re-shaping of the landscape on a miniature scale may be quite feasible when the magnitude of attempting it in the real world would render it out of the question.

I was never a great miniaturist myself, nor do I think there is one simple explanation of this, but it has occurred to me that one reason may be that what I have called the Inner Bailey at Stibbard was large enough to permit quite ambitious projects of civil engineering on a large scale without totally ruining the more obtrusive parts of the garden. Why build a diminutive model house if you can make something like the Fort or the Mine or the tree-house in the yew hedge, big enough to play inside?

By the time Richard reached the equivalent age we were living in the house in Golf Links Road where the garden was far too small to accommodate structures of that size. True, he did make a den on top of the air raid shelter, a not very decorative legacy of the Second World War, useful for storing things in, and certainly providing a firm concrete foundation for a house. However, his inventive genius was equal to the challenge and a complex system of bus routes served every corner of the garden. 'Dinky' buses, some three or four inches in length, threaded their way between the cabbages along well-engineered roadlets from which the surface run-off was conducted through an equally complex system of drains and sewers, all instrumentally graded, into the ditch in the neighbouring field. The miniature bus-routes also permeated the house. Some of the bus stops bore toponymic names deriving from the floor-covering or the furniture, like Brown Lines and Cupboard Carpet; others seemed to be more arbitrary but also more romantic. My favourite, I remember, was Heart of Stickton.

Before Charles had reached this stage in his development we had moved to the house in Hull Road with a much larger, half-walled garden well able to accommodate full-sized earthworks. With his friend Peter Harris he constructed a half-underground den, the only entrance to which was a fifteen-foot-long tunnel.

One of the most common subjects for modelling is the railway. I shall have more to say later about the important role which railways have played in my own world-building, but the actual making of model rolling-stock, model layouts and ancillary landscapes has never been a serious occupation for me. Part of the reason, I suppose, is that I didn't ever acquire the sort of manual dexterity which made my father the handyman *par excellence*. I greatly enjoy the enterprising railway creations with which some of my friends (mostly fellow-geographers, incidentally) have filled their attics, and I can happily spend the odd hour playing with them, but the urge to make them myself has never proved strong enough to overcome misgivings about the cost and anxieties about technical incompetence. So I remain an admirer of other people's handiwork, and for an outlet for my own creative energies I have to look elsewhere.

Neither do I possess the skill or patience to create, as some do, miniature worlds from growing plants. There is a committed fraternity of enthusiasts for whom the *bonsai* presents opportunities to play at sylviculture if not forestry when the real thing would be rendered impossible for most people by the amount of space required. Some find satisfaction in making miniature alpine gardens using real if Lilliputian plants to make a plausible imitation of larger-scale landscapes.

For those of us who, whatever the reason, don't find outlets for our creative urges in the making of physical replicas whether of railways, forests, gardens or anything else, those mental images of reality which we have already encountered in these pages offer innumerable options. We can take over as ready-made products those fictional worlds which have been put together in the imaginations of others, beginning in childhood with tales of giants and fairies and graduating to space fiction; or we can take our raw material more directly from the real world around us. In either case we have the option of making the world subject to, or independent of, the laws of the physical universe to whatever extent we choose. This is what all famous story-tellers have done.

I remember being invited, as will many members of Christ Church, to have tea with the late R. H. Dundas, an eccentric Senior Member of the college, who lived in the corner of Tom Quad in rooms which had once been occupied by a mathematics don, Charles Dodgson. It was a memorable experience to make toast on a fire surrounded by a veritable picture gallery of decorative ceramic tiles on which were depicted various creatures clearly recognisable as the prototypes of some of the best-known characters in the literature of make-believe. Dodgson, of course, was better known as Lewis Carroll, and there in front of us were the original pictures from which his fertile imagination had created many of the personalities of *Alice in Wonderland*, distracting our attention and watching us, in consequence, burn the toast! Very few people have had the inspiration to create worlds like that of Alice, but there must be very many who have attempted the same sort of exercise albeit with less success.

Some people find a greater satisfaction or a more realistic expectation of success by drawing their raw material from other systems of order, such as may be found in mathematics, music or abstract symbolism of various kinds. For such people it may not be necessary to translate the constructs of their imaginations into visual form in order for them to be pleasurable; but I have long recognised that I don't belong to this brotherhood. Visual images have a central place in the world of my own creation. The crucible in which they are made is the imagination. There many of them perish. Of those which survive a few find a mode of expression which allows them to be communicated to others. A great artist would probably see such expressions as the central activity, perhaps even the central purpose, of his or her life. For an ordinary person like me they tend to be relegated to the role of leisure pursuits, peripheral pastimes, fringe interests, but even so the demand for these worlds of the imagination to be translated into a material existence may be very insistent.

It may not surprise you to learn that one of my earliest creative acts was the making of 'places'. My precocious acquaintance with maps allowed me to express imaginary concepts in this form. The first sizeable tract of territory I created was an island with the not very imaginative name of Fordland. Though its latitude and longitude were never determined, its interior details were mapped with great precision on a variety of scales. It was hilly in the centre, fringed with coastal plains and had a heavily indented coastline, and it was criss-crossed by a network of roads and railways plausibly adjusted to the lie of the land. Unfortunately no maps are extant and the memory is too patchy to permit a reconstruction. Like Atlantis beneath the waves, it has long ago vanished.

I suspect its demise may have been hastened by a new challenge which came from a fellow-pupil at the Glebe House, one Richard E Dent. Richard had a very similar type of imagination which he brought to bear on a similar range of interests such as maps, railways and the geography of North Norfolk which was home for him as well as for me. He lived in Aylmerton, a small village which lay two or three miles inland from the coast near Sheringham and Cromer, barely twenty miles from Stibbard. It consisted of a small nucleus with a scatter of outlying farms and cottages. It was somewhat smaller than Stibbard, having a population in 1931 of 311 as against 380, and its opportunities for expansion lay wholly within the imagination of the said Richard Dent, yet, within the limits of that imagination, expand it did. Maps began to appear which showed it to be a very substantial place indeed; so, needless to say, Stibbard had to expand *pari passu.*

The time came when Richard and I went on to different schools, but we continued to develop our respective town plans which we mailed to each other for at least another two or three years. All the plans of Stibbard, therefore, finished up in the possession of Richard, who, like so many of my friends, was destined not to survive the war. It may be assumed that *The Stibbard Collection* has not survived either, and for a long time I feared the same could be said of the Aylmerton papers; but a few years ago I discovered a *Map of, and Guide to the City of Aylmerton, Norfolk,* so, while I can't illustrate the Stibbard which grew out of my imagination, I can reproduce details (Figures 33 and 34) which had simultaneously grown out of his. The original is very tatty, executed in the blue-black ink of a fountain pen typical of the nineteen thirties on ruled foolscap paper folded to make an eight-page brochure, but it is a labour of immense love and

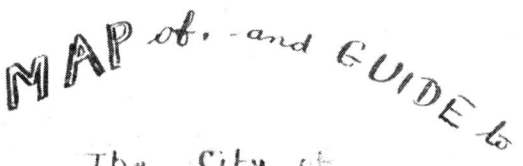

FIGURE 33. MAP OF, AND GUIDE TO THE CITY OF AYLMERTON, NORFOLK.
Cover of an 8-page brochure, blue-black ink on ruled foolscap,
by the late Richard E. Dent, *circa* 1935.

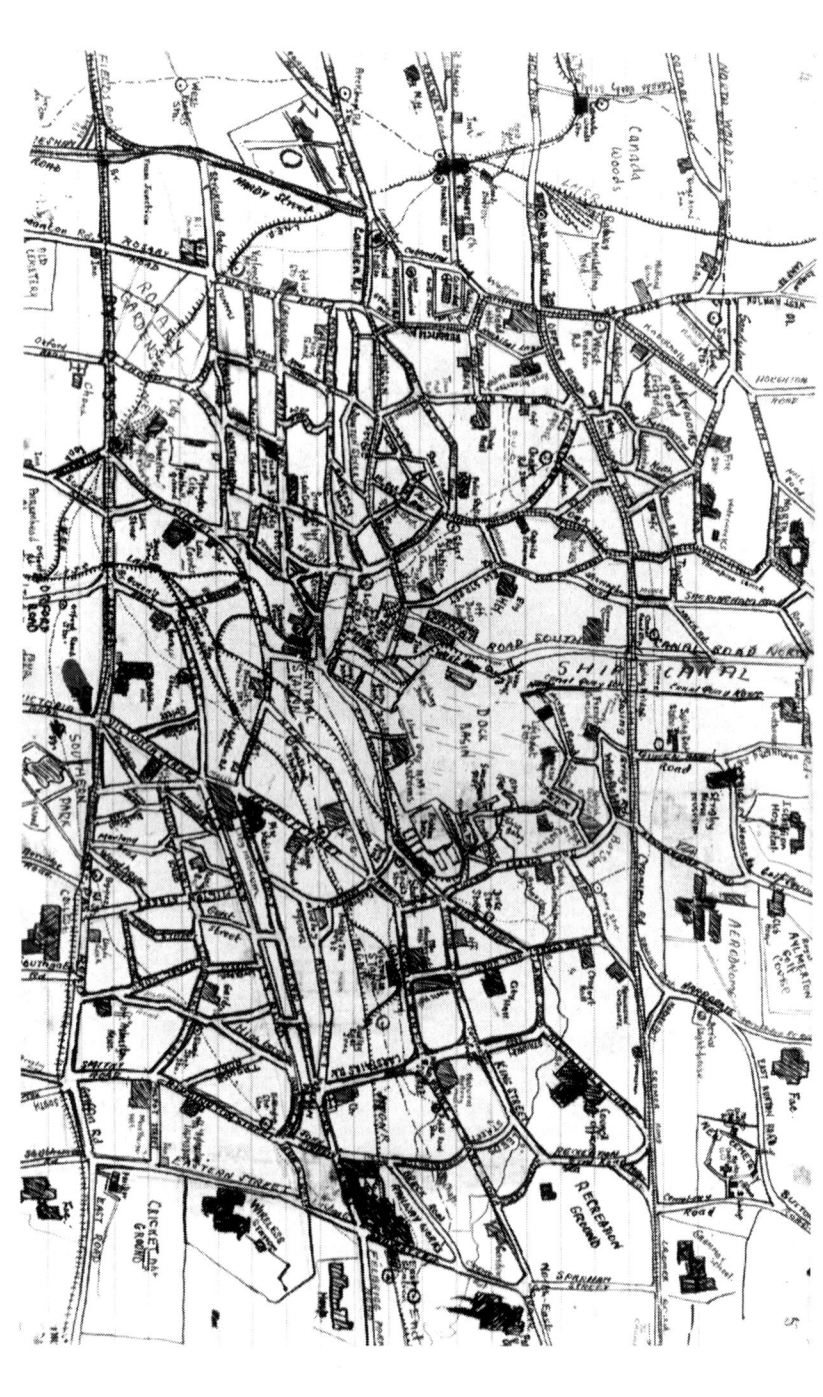

FIGURE 34. PART OF THE STREET-PLAN OF AYLMERTON, NORFOLK. Note the compliment implied in 'Stibbard Road' which can be seen halfway between the centre of the map and the left-hand margin. Detail from 'Map of the City of Aylmerton'. (See Figure 33.)

the only tangible souvenir I possess of a good friend and a kindred spirit.

I dare say that, since most of these maps and guides were produced during my early years at Shrewsbury, they can be interpreted as visible signs of the imprisoned geographer urgently seeking a means of expression which was not provided in the academic curriculum. The imagination is a processor of ideas. It has an input and an output, and between the two something happens to turn the former into the latter. No doubt my teachers were hoping that, by pumping in Cicero and Thucydides at one end, they would ensure the issue of some really worth-while product at the other. But the classical authors had no monopolistic protection against competition. They had to share the inlet valve with much other material of my own choosing, and it was the influence of the Midland & Great Northern Railway and the physical geography of North Norfolk which provided the framework of the Master Plan of Stibbard, as also of Aylmerton. The idea of two such tiny villages burgeoning into the sort of city depicted in Richard's map must have struck us as plainly fantastic; yet, come to think of it, they were each recorded in the 1931 Census as being more than twice as large as that tiny village of 154 souls in Northern Buckinghamshire which went by the name of Milton Keynes.

If I had known at the age of eighteen what I know now I should probably have aimed at becoming a landscape architect. The idea of manipulating the real environment to create a real world out of one's own imagination has an immense appeal; but the profession was at that time very small, and I doubt whether I even knew of its existence until much later. I doubt also whether I would have been very successful, since the criterion of excellence is the satisfaction of the client, and I suspect my own tastes would too often have needed to be suppressed in deference to the dictates of fashion. In fact, the only exercise in landscape design I have ever had the opportunity of carrying out is a small corner of my own garden with the aid of cypresses and other plant material now generally regarded as the hallmark of bad taste (Figure 35).

Most of my creative activity, therefore, lay in the field of writing, not least about landscape, and I have derived much satisfaction from this; but writing, whether in prose or in verse, is an intermediate vehicle for conveying landscape images. Although it can convey impressions which direct visual observation cannot, it is also limited in its capacity to communicate an accurate and realistic picture of what places look like in detail. The literary and the graphic arts are complementary, and I often regretted that I had not acquired more than the most rudimentary skills in drawing, painting or other media of visual representation. My little granddaughter, another Helen

Mary Appleton, Richard's daughter, has more competence in painting at the age of eight than I had at thirty. Perhaps she gets it from her mother, Julie, whose father was an art teacher.

Anyhow, better late than never! When I was nearly fifty Iris persuaded me to attend an evening class in oil painting. She was involved in running a nursery school with a group of friends, one of whom, Jill Williams, was an artist and art teacher, and it was she and her successor, Jane Gear, who taught me the elements of landscape painting in a building which went by the name of Cottingham High School by day and Cottingham Evening Institute by night. My competence has never progressed beyond the mediocre, but I have mastered the simpler techniques enough to be able to introduce into a painting something out of the imagination which goes beyond the mere reproduction of a photographic likeness. A fairly extreme example is illustrated in Figure 36. I painted it on my return from a holiday in the Cyclades Islands and the building style owes much to the local architecture, particularly that of Santorini; but the composition comes entirely from the imagination and seeks to give expression to some of my theoretical ideas in landscape aesthetics to which further reference will be made later especially in Chapter 9. These ideas, however, were formulated even later than my initiation into the art of painting, and before we examine them we must look at some other broad topics which have a bearing on the evolution of my own habits of perception, spinning a few threads, as it were, to weave into a more coherent tapestry later.

Taste

Of all the phenomena that lie within the human experience one of the most difficult to handle is 'taste'. There are two basic problems. The first is that the word itself has a whole range of different meanings. In its simplest meaning taste is one of the five senses, along with sight, hearing, smell and touch, but even here complications arise. Taste, strictly speaking, is that sense by which we are able to distinguish between such qualities as sweetness, saltiness, bitterness, etc.; but when we talk about tasting food, almost always we are referring to a complexity of sensations in which smell is at least as important as taste in its strictest sense. A wine 'taster' will use words like 'bouquet', even 'nose', words which are more properly concerned with the sense of smell.

From this more limited meaning the word is extended to cover other sensory experiences, hearing, for instance, (a taste in music), seeing (a taste in wallpaper, pictures, landscape), and from there to embrace more complex experiences like a taste in theatre. Very soon, also, we encounter the assumption that taste is not just different as

FIGURE 35. 'THE ONLY EXERCISE IN LANDSCAPE DESIGN . . .'
Personal preferences prevail over the fashions approved by the pundits. There have been some deliberate attempts to apply the principles discussed in Chapter 9 to this little corner of the garden at 39 Hull Road, Cottingham. Photo by the author, 1987.

FIGURE 36. CYCLADIC INVENTION IN TINACHROME.
An imaginary composition using architectural styles from the Cyclades Islands to
illustrate the principles of prospect, refuge, etc., (See Chapter 9). Painting in oils by
Jay Appleton, reproduced by permission of the owner, Tina Harrison, to whose taste
in colours the arcane title is intended as a compliment!

between one person and another, but that it is 'better' or 'worse'. We speak of 'good taste' and 'bad taste', thereby introducing moral concepts into our distinctions.

The second problem is that this very diversity between the tastes of different people inevitably raises questions about the rules, if any, which operate. As long ago as 1757 the great Irish politician and philosopher, Edmund Burke, put it like this:

> . . . If Taste has no fixed principles, if the imagination is not affected according to some invariable and certain laws, our labour is like to be employed to very little purpose. . . (Burke, 1958 edition, p.12).

There is a sense in which tasting is much the same as testing, a word to which it is etymologically related. Tasting food is one way of making sure that it's acceptable, and in all the applications to which the word lends itself the idea of experimentation is never far away. We like things we're used to, but we also like trying new experiences provided they don't involve innovations so radical as to cause us to feel threatened. We all vary, not only in our tolerance of novelty, but also in the different areas of experience in which we enjoy it or fear it. In my own case I recognise that my taste is in some respects exploratory, experimental and aimed at achieving new sensations, encountering new situations and learning new facts about the world around me. In other respects it has been so conservative as to cause dismay among my friends and relations, and nowhere more so than in the matter of food.

The fact is that my taste in food was very slow to change, and even after I had grown up I still liked things as plain and simple as possible. It didn't worry me at all if I had to eat the same things day after day as long as they were things I liked. I must have been the most unadventurous eater of anyone I know. When I heard a dish described as 'interesting' my heart would sink; so I was left with a very limited range of dishes which I really enjoyed. I can only describe my taste in food as infantile, and, although it has by now broadened considerably, I still prefer fruit jelly to caviar or oysters! The meat-and-two-veg. on which I was brought up still represents the pinnacle of gastronomic experience and I am constantly amazed at the lengths to which restaurateurs will go in their search for more complicated, more expensive and more time-consuming ways of rendering good food uneatable. There are some foreign dishes which I enjoy very much, but, generally speaking, if I were to make a list of my preferences, it would still be absurdly xenophobic. Needless to say, I'm not proud of this; on the contrary, I wish I could eat and enjoy anything as my father could and, with few exceptions, my children can, but it hasn't worked out like that.

So why not?

I was by nature a rather timid child. I moved cautiously from the discovery of one situation to the exploration of the next. In my investigations of my external environment, as represented by the hedges and ditches of Stibbard, the woods, the streams and everything else that comprised the landscapes around me, I seem to have been precociously efficient at making new discoveries and fitting them into my developing framework of experience and understanding. Self-confidence tends to grow out of success, and since most exploratory adventures were successful they conditioned me to face the next challenge with pleasurable anticipation. There were, of course, exceptions. When my mother told me that, as a very small child, I had been frightened by a dog and by some practical joker, who should have known better, bursting a balloon in my ear I began to understand why I was so ultra-cautious towards dogs and why I so disliked, and still dislike, balloons. Usually, however, curiosity has led me into adventure, and consequently the exploration of new places has been one of the most pleasurable experiences in my life.

How different was the exploration of new kinds of food! It may be that, by gentle encouragement, I could have been gradually induced to widen my tastes, to move forward in my own good time, slowly familiarising myself with one sensation so that I was ready to move on to the discovery of the next. I think now that the experience of being forced to eat up all my meals at the age of nine and ten finally killed off any self-confidence I might have been acquiring in my own capacity to master this particular kind of exploratory adventure. The mouth is one of the most highly sensitive and personal parts of the human anatomy. To force a child to admit a foreign body into it against its will is nothing less than an act of oral rape, and like all acts of rape it may well cause a life-long aversion to any situation which conjures up an association, consciously or unconsciously, with the original outrage.

Since that time I have rarely found myself actually wanting to try anything new unless I have reason to believe that it closely resembles something I already know I enjoy, and such tastes as I have subsequently acquired I have been introduced to by very gradual processes subtly engineered by an understanding and diplomatic wife. If there is any area of my life in which I can see the inhibiting influence of childhood experience on my developing tastes, this is it. I was unwittingly but systematically driven to a kind of gastronomic agoraphobia.

The desire for simplicity and familiarity in food also finds expression in a strong preference for a binary classification of dishes into two categories, savoury and sweet. As long as I can place a tasting experience firmly in one category or the other I know where I am. Even long-established practices, which involve confusing the two, worry me. For instance, I'm very fond of roast pork and also of apple sauce, but I invariably eat one before the other. The same applies to mutton and red currant jelly, turkey and cranberry sauce, and so on. I remember as quite a small boy teasing my father about 'putting jam on meat'. When, after nearly thirty years of exile, I returned to live in Yorkshire and encountered the local practice of eating cheese with apple, I would again enjoy them both consecutively rather than ruin them both by eating them concurrently. It follows that many of the innovatory ideas which have more recently found their way into English cuisine, principally from various foreign traditions, and which seek to exploit every possible combination of tastes, seem to me retrogressive. 'Sweet and sour' threatens the very basis of my alimentary preference system.

In laying before you this rather pathetic account of my whims and caprices - all right then, fads - in that area of experience which bears the most literal interpretation of the word 'taste', my intention is to establish the simple fact that it's quite possible for one individual to find complete satisfaction within a range of experience so limited that others would find it monotonous, unadventurous and just plain boring. *Per contra*, it may be that people who are contemptuous of what to them is a deplorable lack of enterprise in eating habits are themselves unexcited by the sort of exploratory adventures which I would find most stimulating. I hope you'll bear this in mind if you find, as well you may, that those idiosyncrasies which I'm about to describe, and which have fascinated me so deeply, are quite incomprehensible to you! Anyone who is conscious of failure, inadequacy or social disapproval of his or her achievements in one field is naturally inclined to attach more importance to another in which success has come more easily, bringing with it satisfaction and further encouragement, so my precocity in what I shall call 'orientation' came to my rescue as a compensation for my inability to keep up with my peers as a *connoisseur* of 'good' food.

Distance and Direction

I have already referred more than once to my early interest in understanding the geographical inter-relationships between the various components of the environment. The usual device for representing spatial relationships is a map, and the conceptual image which we build up in our own minds is called a mental map, though the phrase has only come into common use in recent years. In building up such a mental map we rely heavily on abstracting information from our own visual experiences. The great majority of these, however, consist of horizontal or near-horizontal views projected on to the retina of the eye by rays of light reflected from the surfaces of objects lying at varying distances from the point of observation and subtending different angles between each other at the point where they meet in the eye. If we can accurately calculate these distances and these angles we can determine how these objects are distributed on the map in relation not only to the point of observation but also to each other; but in order to do this we need to be able to convert the original observed image from elevation to plan, and many people, even quite well-travelled adults, seem to have great difficulty in doing this. I remember grown-ups being greatly surprised to discover that, while I was still quite a small child, I could do it better than they, as long as the images themselves derived from my own personal observations.

As soon as I had to rely on information from other sources problems arose! Once the element of reality, of actuality, of immediate experience was lost, errors of gigantic absurdity distorted the emerging mental map. Even when I had reached the stage when I could name all the countries of South America I remember thinking how exciting it must be to stand on the top of the Andes in Bolivia or Peru and to look westward to the Pacific Ocean, almost at one's feet, and eastward to the Atlantic which I envisaged as a distant line of blue, just visible beyond the vast expanse of the forested Amazon Basin! Had I paused to work it out I should have known with certainty that the field of visibility would reach only a tiny fraction of the way to that distant shore, but the reality from which the image was derived was the reality of the atlas and not of direct observational experience, and it just wasn't 'real' enough to provide the basis for an accurate conversion, in this case from plan to elevation.

Just as errors can creep into one's assessment of distance, so also of direction. An early familiarity with the convention which places north at the top of a map suggested to me some absolute reality in the points of the compass, and this reality could be checked against common experience, at least on a fine day, by the passage of the sun through a predictable sequence of directions. To anyone with a keen visual sense the idea of east and west, as the directions in which to look for the rising and the setting of the sun respectively, is as real as the idea of near and far or of up and down.

Since my own early introduction to the points of the compass provided me with such a powerful sense of direction, or perhaps it would be more correct to say 'awareness' or 'feeling' of direction, it seems strange that, in the very cradle in which it was born, I should have got it wrong! I mentioned earlier that the orientation of the house at Stibbard was not due north-south; in fact it deviated by about twenty-two degrees from true north. To make it fit tidily it would have been necessary to move the points of the compass by that amount in a clockwise direction. In my own little mind, however, I 'corrected' the orientation by, as it were, moving them the other way, anti-clockwise, so that the error was not twenty-two degrees but sixty-eight! I imagined the front door to be facing north when in fact it faced west-north-west, and all the other directions at Stibbard rectory were equally distorted.

The misconception must have dated from a very early age because I was able to correct it rationally as soon as I understood about the movement of the sun, but the erroneous concept was so deeply ingrained that for years afterwards the initial, spontaneous image remained wrong and always needed to be corrected by a conscious effort. At the age of sixty-five I visited the village again for the first time in decades and was momentarily surprised to see the sun in what I subconsciously thought was the south-west at ten o'clock in the morning!

Many other images have been fed into this distorted framework. I referred earlier to my introduction to the *Rubáiyát of Omar Khayyám* which I can date within a year to the age of twelve, long after I had logically corrected my directional misconception, yet the picture I had (and still have) is of the Hunter of the East catching the Sultan's turret, which for some reason stood in Stibbard churchyard, in a noose of light cast, as it were, from the north.

Much later I had another experience which brought home to me even more strongly how deeply ingrained our concepts of direction can be. At the age of forty-one I lived for about a year in New South Wales. I was, of course, fully prepared to see the sun pass through the northern instead of the southern sky and I frequently observed it doing just that; but years later, recollecting in England my experiences of those years, I often imagine the house we were living in casting a shadow on the patio which lay on the *northern* side, exactly as it would have done in the northern hemisphere but couldn't possibly have done in Australia.

In spite of the enormous importance to me of direct visual experience, here is a case of the conceptual framework into which I poured that experience proving even stronger - more real than reality! The story used to be told of an old and obstinate Cambridge don (an Oxford don if the story-teller was from Cambridge) who, on being asked what would happen if the theory didn't accord with the facts, replied 'So much the worse for the facts!' It's a strange sensation indeed to discover that this is borne out by one's own experience in precisely that field in which one believes oneself to be most keenly aware! It should lead you, if you're of a critical turn of mind, to discount everything I say in this book, since my testimony is clearly untrustworthy; but I hope you will instead charitably recollect some of those fallacies which must surely have insinuated themselves into your own assessment of the world around you, and if you tell me there are none it will be my turn to disbelieve!

The convention of placing north at the top of the map, which eventually became generally accepted in cartography, has the great advantage that it makes comparison relatively easy. Yet this is not the natural way to transfer an impression of landscape from elevation (the form in which it is pictorially perceived) to plan. If we stand on a tall tower and look out in one direction the ground immediately below us is observed vertically, the extreme distance horizontally. The ground at the base of the tower is perceived in plan, the distance in elevation. Between the two extremities there is a gradual transition. If we take a piece of paper and draw at the bottom the ground which lies at the base of the tower, we have made, to all intents and purposes, a map. Objects which lie further away from the base of the tower will occupy progressively higher positions on the paper until, if we have room, we will reach the horizon at the top. People who draw sketch-maps of how to get to places frequently follow this procedure. The bottom of the map is equated with the point of origin of the journey; the top is the destination. In other words the orientation of the map assumes that we are looking from the origin to the destination.

In trying to establish the orientation of pictures of remote places which lay beyond the reach of my own direct observation I would employ one of three principles. If I had a vague idea of where the place was, say in China or Africa, I would probably have a mental image of it on the world map with north at the top. Applying the principle I have just described, the horizon in the picture would therefore 'read' as if it were in the north. To put it in another way, the viewpoint would be assumed to be from the south.

The second principle applied if I imagined that I was looking at the place directly from home. Thus I always imagined that the view of the bell tower in Bruges, which had so impressed me in *The Outline of the World Today*, was from the north-west, since I assumed (almost but not quite correctly) that Belgium lay to the south-east. Not until 1985, when I took the photograph reproduced as Figure 37, did I find the actual viewpoint and discover that it was precisely the opposite of what I had imagined.

Yet a third principle could apply if the actual direction of the view was known at the outset. Consider, for example, the well-known view of Capetown with Table Mountain as seen from the sea. Applying the first principle, the map of South Africa would come to mind, north should be at the top and England should be somewhere

FIGURE 37. THE BELL TOWER, BRUGES, BELGIUM.
View to the north-west. Photo by the author, 1985.

beyond the horizon; in other words the view should be from the south. If the second principle applied, South Africa should be in the south (from Stibbard) and therefore the view should be interpreted as from the north. In fact I early recognised it as from the west (correctly), probably because the conspicuous details of the picture - mountain, city and sea - could be identified on the map to which my mother had doubtless matched it. Here then, the third principle applied.

For some reason, however, all pictures of Washington D.C. appeared to be taken from the east (applying the second principle), because I imagined myself looking west towards North America. It is interesting to speculate that, had I been raised in Cincinnati or San Francisco, all such pictures might have appeared to be taken from the west; but I wasn't, and therefore almost all cities on the North American continent appeared to be seen from the east.

I say 'almost all' because there were exceptions, in which one of the other two principles applied. For example, I referred earlier to a picture, which made a strong impression on me, of the Canadian Houses of Parliament in Ottawa (p. 48). I think I must have known, under my mother's tutelage, that the Ottawa River ran roughly from west to east - that much would have been apparent from the atlas - but I suspect the scale would have been too small to establish clearly on which side of the river the city lay, or perhaps I had read it too casually to notice. If the river ran west-east it must follow that the façade of the building which crowns the sloping river-bank must face either north or south. (The idea that the river might wind was too complicated to worry about.) To settle this I would appear to have invoked the first principle, that is to say, if the top of the picture corresponded with the top of the map, i.e. the north, the view must be in that direction and the building must have stood on the north bank of the river. A more careful subsequent perusal of the map had established the truth long before I first visited Canada at the age of fifty-two, but it still came as something of a shock to see the Parliament Building for the first time occupying a position every bit as magnificent as I had expected but on the wrong side of the river! Not only that, but every time I see a picture of the Parliament Buildings, as seen from the river, I have to make a conscious, rational act of re-orientation to get it right, although I have visited it several times. No such problem exists with pictures taken from the other side, which is in fact the front of the building, presumably because no spontaneous connection is made with the picture which sparked off the original mistake.

Mistakes of this kind still occur, most commonly, perhaps, when I arrive in an unfamiliar place after dark and make a wrong initial appraisal of my orientation. Let me cite a recent example. A few years ago I visited for the first time the island of Bali in Indonesia. I arrived at the airport after dark on a cloudy evening (which precluded the use of the stars to maintain a sense of direction) and was taken by car to a 'village hotel' consisting of small individual cabins in a coconut grove. After a very circuitous journey, calling at a number of intermediate stops, the driver made a right turn into the hotel grounds. Having previously perused the map, I expected to arrive from the south, in which case a right turn would take us eastwards; but we had described so many twists and turns that I had entirely lost my sense of direction and didn't realise that we were actually approaching from the north. Since I knew that the hotel was on the east coast of the island I therefore assumed that it lay between the coast road and the sea.

I discovered my mistake as soon as the sun rose the next morning. Every day for the next week I crossed the road to go to the east-facing beach. Every day I looked across the Badung Strait to the island of Nusa Penida which I knew from the map to be in the east. Every day I was mesmerised by the volcanic cone of Mount Agung, which I equally well knew to be in the north-east. On three evenings I photographed the spectacular sunsets which made an unforgettable impression. About the points of the compass not a vestige of intellectual doubt remained, but I could never quite banish the *feeling* that everything was the other way round! It required, and still requires, quite a laborious intellectual effort to over-ride the gut feeling and to correct an illusion the origin of which was quite different from my misconceptions, referred to earlier, when I lived in New South Wales. There I had my orientation correct from the start. Only in retrospect did I mis-remember the position of the midday sun and erroneously make it accord with the more familiar situation in the northern hemisphere. As for my disorientation in Bali, had I first arrived on a sunny day, the misconception could never have occurred and my recollection of that hotel would have been quite different.

I have no doubt that some readers will be able to match these anecdotes with others drawn from their own recollections, but that many, and probably most, will find the whole idea of constructing directional frameworks, into which landscape images may be fitted, entirely foreign to their own experience. To them the direction from which a landscape appears to be viewed is wholly irrelevant to the sensation it conveys. To me, however, every landscape is like a detail from a larger painting, a part selected and artificially detached from a more extensive whole.

I think this ties in with my early awareness that an horizon is an arbitrary line of separation, dividing the landscape in front of it, which can be perceived by the eye, from the landscape beyond it, which can be reached only by the imagination. The view from my bedroom window at Stibbard was like that. The River Wensum was a part of the view, even though it lay a good mile beyond the horizon out of sight. The view of the western Pennine edge as seen from Bradley Hall was a wholly Lancashire landscape, but its 'feel' encompassed Yorkshire, which lay beyond it, as well. The whole picture was a kaleidoscope of images, partly the subject of observation, partly the creation of the imagination, largely erroneous but very real, which seemed to hang, as it were, out of sight, suspended from an infinitely distant sky.

When, therefore, having been shown a photograph of a painting of a place, and, having initially assumed the aspect to be in a particular direction, I later discover I've been wrong, the picture itself looks quite different, because it fails to fit into that wider geographical framework which contains not only the view itself but also what the imagination has supplied, beyond the reach of the eye, which to me is an integral and important part of the *Gestalt*. Many studies have been made of landscape preference, based almost entirely on reactions to landscapes as depicted in photographs, and there is much evidence that people's responses to two-dimensional photographs differ little from their responses to three-dimensional observations made in actual landscapes (Kaplan and Kaplan, 1989, pp. 16, 17 and 39). I accept the probability that this is generally correct; but I can't help wondering whether this 'reading' of imagined directions into photographs doesn't make me an exception to the rule. Even landscape paintings of imaginary places I invariably perceive as if viewed from some particular point of the compass. If then, by an act of will, I imagine them as viewed from another direction, they become almost like different landscapes; but I don't expect this to be a common sensation shared by everyone.

Be that as it may, there are within everybody's experience innumerable kinds of interaction between the two worlds, that of 'reality' and that of the imagination. The latter is created by re-processing raw material drawn from our observations of the former, material which has already been re-processed in the act of observation itself in a highly selective way. The re-processing takes place in a number of stages at any of which the initial image may be significantly changed. Any deficiency in our physical capability to observe our environment, failure of eyesight, for example, will introduce a distortion of the original image. Then we may fail to make a proper appraisal of the significance of what we have observed for our own needs and purposes. We vary greatly in our ability to make such appraisals, but, having done our best, we must transfer the important information to the memory, otherwise we shall lose it and it won't be available when we need to act on it. We don't need to be reminded, either, that memory can let us down! And so it goes on.

Remembering and Forgetting

When we come to draw on those fragments of previous experience which are still stored within the memory, a highly distorting process has already been in operation and any world which the imagination can build out of such fragments will differ widely from the original while retaining tenuous links with it. Each of us has a past which can never be wholly erased (Lowenthal, 1985).

The extent to which my own memory has proved to be highly selective was shown up recently when I re-visited Stibbard after a very long absence. The village had, of course, changed. New houses had been built; old cottages had been pulled down. Little willow saplings, which I well remember being planted, blocked out the view of the hedge-munching trees on the turnpike. The Rectory garden, now the Old Rectory garden, had been greatly changed. The denser parts of the shrubberies had been cleared out. The beech trees which had towered over the house and even over the church tower had gone; the pink chestnut tree too. The whole of the garden beyond the wall had been sold off separately leaving the churchyard as the only part of the Outer Bailey to survive. The churchyard admittedly had changed very little apart from accommodating more tombstones, many of which bore the names of people who had been my friends.

There was much in the village also which had not changed. Some of it was exactly as I remembered it, while other parts had faded from the memory. I found myself surprised to come across field gates, barns, even cottages the very existence of which I had entirely forgotten, yet not entirely, because, when I began to think about the gates I remembered climbing over them; when I looked up at the barn roofs I remembered scrambling up them, and when I put my mind to it I could even remember who had lived in the cottages.

Some of these nooks and corners I had called to mind during the years between. Sometimes they had been changed in the very process of being recollected; sometimes they remained exactly as they had been. The mental map of Stibbard had been updated many times, though by the imagination only and not by direct contact with the original, and, when I began to think about it, I realised that the same thing must also have happened to the other models which I had been carrying round in my head. Bradley Hall, which I re-visited at about the same time, after an even longer interval, had changed much more. It had been taken over during the war and, with its garden and the neighbouring field where I had watched the trains, converted into a munitions factory and afterwards used by a food manufacturer.

Today Bradley Hall is a restaurant. The lawn and the weeping ash are still there but the surrounding rhododendrons have largely disappeared and the peripheral hedge which so effectively blotted out the world beyond is so thin as to be almost transparent. It is as though the whole system of protective defences on the western side, on which the mood of sweet enclosure had depended, had been dismantled. There is no more real privacy. Half way down the drive a railway cutting passes underneath, a legacy of its wartime function. To the south my grandmother's beloved rose garden has a covering of tarmac. It makes a convenient and capacious car-park for the patrons, but thank God my grandmother never lived to see it! Between the house and the railway what used to be my favourite meadow is full of small industrial buildings.

If memory failed here there would be no possible means of resurrecting the finer details of the house and garden I knew so well. My recollections of the surrounding lanes and fields are indeed hazy, but the image of the Hall itself, its outbuildings, its rhododendrons, the contrasting smells of the old coach-house, where Uncle Henry kept his car, and of the little black shed beside Uncle Edge's lethal railway line - all these remain vividly in my head. Vividness and accuracy are, of course, very different things, and how reliable a record it is of what stood there fifty or sixty years ago I can't say, because the prototype is no longer there for comparison. This was precisely the problem encountered by the poet, John Clare, when the open fields of his native Helpston had been made unrecognisable by the hedges of the Enclosures.

Places which I encountered in childhood only briefly have been even more susceptible to alteration in the course of re-processing, because the original pictures were less firmly fixed. I seem to share what I believe is a common experience among people who are growing older, namely an increasing tendency to see these early environmental images through rose-coloured spectacles. The landscapes of childhood take on something of that idyllic quality which distinguishes, say, the paintings of Claude from the landscapes of Peninsular Italy as seen by today's tourists. Looking into one's own imagination is rather like going into a picture gallery. There one sees representations of the same world which one has left outside, the same yet so different.

The World of Dreams

Whether or not dreaming has some useful, practical value in helping us to cope with the problems which confront us in our waking hours, as is generally believed by most psychologists today, it can surely tell us something about our own habits of picture building, so I want to spend a little time now looking more closely at this very special kind of re-processing of raw material drawn ultimately from our observations of the world around us.

Many people have repetitive dreams, and I'm no exception. One of the earliest I can remember is the Three Moons dream. I was travelling eastward, or possibly north-eastward, in the side-car of my father's motor-cycle. Even at this early period the direction of travel was an integral part of the scene. The road was unfenced and over on the right there stretched out a broad expanse of open grassland sloping gently upwards to a fringing belt of pine trees perhaps half a mile away. From the horizon ahead the moon rose slowly but quite perceptibly above the pine trees, to be followed by a second and, after an equal interval, a third. Having passed over the pine trees the moons disappeared behind me.

How often I had this dream I can't say, neither can I hazard a guess at its symbolic meaning, if it had one. What was important to me was the moonlit landscape and the recognition that there was something in it which triggered off a deep emotional response. It was probably the earliest event in which I can clearly remember experiencing that powerful kind of attraction towards a landscape which I was to feel many times in the future, much more than a mere pleasant feeling, something more akin to falling in love. I wanted desperately to believe that this place really existed and, if possible, to find it, and yet I was afraid that if I ever did it would lose its magical touch. Within the last few years it has occurred to me that I may have been taken on a moonlight night at a very early age across the Breckland of south-west Norfolk. This is little more than twenty miles from Stibbard on the road to London, and resembles my dream landscape more closely than any I can remember seeing subsequently, though I certainly couldn't identify a particular spot as the origin of the fantasy. It could even have come from a picture book. Why the three moons? Don't ask me?

The Three Moons dream belongs to the distant past. More recently I've had other recurring dreams which seem to say something about attitudes to landscape. One of these features a crooked, irregular street flanked by irregularly grouped semi-derelict houses and other nondescript buildings overhung by tall trees. I'm either in a car or on foot, changing from one to the other as one is allowed to do in dreams, through various environments I've never seen before, when suddenly I find myself in this place. I know my way now and can pick a route through it, always travelling westward, partly in the street and partly through the buildings. A small stream crosses the street in a ford, and beyond it the track rises up a hill and disappears over a brow. As I complete the transit I'm aware of mixed feelings of attraction towards the place and of a desire to escape from it as soon as possible. Safety lies beyond the brow of the hill, and an awareness that escape is impeded by the climb is an integral part of the act. As I climb the slope I look back. There are usually people there; whether they're hostile I'm not sure, but by the time I see them I know I'm safe. The anxiety begins as soon as I recognise that I'm in this particular dream and it ends as soon as I reach the brow of the hill.

Whatever the psychological significance of these recurring dreams, the important thing for our present purposes is that they have been sufficiently frequent and repetitive to establish the same sort of familiarity with 'dream places' as one commonly establishes with places in the real world. The details of an imaginary place, in other words, are so clearly committed to memory that it becomes possible to anticipate what one is about to encounter next as one travels through it in exactly the same way as one might in the real world. To all intents and purposes dream places have achieved the status of real places in the mental map.

In addition to repetitive dreams which feature the same recognisable if imaginary places, there are certain repetitive elements which figure in otherwise non-repetitive dreams. For example, I often find myself in a dream travelling on a train when I become aware that I have to get home. This kind of dream differs from what I have just described in that the places themselves are never the same, though I always have a clear awareness of where I am. Quite recently I've had many dreams of this kind. I usually have a vivid impression of the map of Britain, (I don't remember having a dream of this kind which took me abroad), and I'm able to work out the relative advantage of going by different routes; not only that, but I realise on waking that my calculations have generally been quite logical and not altogether out of accord with the real railway map. There is, however, one curious feature of these dreams. 'Home' is *always* Stibbard, even though I haven't lived there, or indeed visited it more than three or four times, for well over half a century. To me this suggests that, if this is the homing instinct, it must be no easy matter to detach it from its original base and plug it into somewhere else.

Many of my dreams have featured fires in the night. I can recollect several incidents from childhood which might have contributed to the stock from which their ingredients were drawn, though I'm not suggesting that there is a direct relationship between any of these incidents and any particular dreams. I remember going out one night in the side-car and being stopped at a level crossing near Thursford, a few miles from home. A train came past with the firebox door open, and not only did the firelight fill the cab of the engine with a pool of light, but it also spilled out into the encircling darkness, staining the overhanging foliage an orange-red. Other vivid recollections of nocturnal fires take me back to Bradley Hall, where more than one ironworks in the vicinity of Wigan used to tip molten slag. The display was particularly effective if there was a cover of low cloud on which a brilliant patch of almost white light would momentarily appear above the slag-heap, rapidly turning through orange to red and gradually fading to return the cloud-canopy to an amorphous darkness.

Of the hundreds of dreams I must have had about nocturnal fires and red skies I remember one particularly well. It was in the late nineteen thirties and I was standing on the tennis court at Stibbard looking towards Norwich. The sky was glowing brightly and I got the feeling that the whole city was ablaze, though at first I couldn't actually see the flames. After a while they appeared like little tongues, and I remember with what relief I woke up to the realisation that it hadn't really happened. I'm not suggesting that it was in any sense a predictive dream, but it could only have been about a couple of years later that I found myself standing on the low hills near the boundary of Cambridgeshire and Bedfordshire looking towards the southern horizon. The night sky was indeed reddened, not this time over Norwich but over London, some fifty miles away. Innumerable columns of smoke were rising into the cloud base, and as they caught the light from the conflagration they seemed to be picking up the fire itself and carrying it aloft into the sky. There was nothing dreamlike about this. It was one of the heaviest air raids of the war.

Mention of Norwich brings me to another category of dreams, those which represent the familiar in an entirely different guise. These resemble the dreams about wholly imaginary places which I described above in that they are recurring dreams, but they differ from them in that the places concerned, though thought of as real places, often in their correct geographical locations, look quite different. Norwich was such a place. In Figure 38 I've attempted to outline its main features. While I find it impossible to draw accurately its architectural details, what is important is the ambience, the general feeling of exposure and enclosure brought about by the disposition of the building blocks and the spaces between them in relation to the lie of the land, and these have been so deeply etched into my memory that I have no difficulty in reproducing them. As soon as I begin to fill in the architectural details, even where I think I can remember them, to make buildings into plausible architectural designs, the picture begins to lose something of its authenticity.

As you can see from the drawing, there is a ridge of higher land which rises above the surrounding plain and on the top there stands the railway station. It's the highest feature of the skyline. On its

FIGURE 38. 'NORWICH' FROM THE WEST. For explanation see text. Pencil sketch by the author.

FIGURE 39. 'PRIVATE EYELET'. The School Buildings, Shrewsbury, as seen in a dream. Cf. Figure 19 and see text for explanation.

western side the ground slopes down, the upper part being occupied by buildings of, perhaps, three or four storeys in height. Viewed from the west these buildings form a kind of back-drop to a large open triangle at the apex of which a passage leads towards the station. I'm aware of another approach to the station round the northern end of these buildings and vaguely conscious that the whole complex is surrounded by some kind of urban development.

Anyone who knows the real Norwich may be tempted to interpret the ridge as the mound on which stands Norwich Castle, supplanted in my version by the railway station in conformity, no doubt, with those obsessive interests which I'll be telling you about in the next chapter. It's true that, in the real Norwich, this mound is flanked on the west by the Market Place with blocks of buildings of about the right height separating the two, but not only the configuration but the whole feel of my open space is quite different. The station, if it had a prototype, could be modelled on the High Level (Great Eastern) station in Cromer, twenty miles to the north, which also stood in an exposed and elevated position with rail access from the south. My 'Norwich' station was quite small for a large city. The platforms were covered by low glass roofs but were quite open at the sides so that one could obtain a wide all-round view.

The site of the real Norwich is separated from the flat land of the Norfolk Broads by a few miles of gently undulating terrain, but from my station-citadel the ground fell away abruptly into a vast plain, at the further end of which lay Great Yarmouth, more or less in the position in which you'll find it on the Ordnance Survey map. At the southern end of the station platforms the railways diverged. The London line headed off to the south while the Yarmouth line curved sharply eastwards and descended immediately to the plain. Every time I had this dream, while the details may have changed considerably, these basic features remained so constant that I could quite easily find my way around.

These Norwich dreams belong to the fairly remote past though they have remained vivid in my memory, but much more recently I have had recurring dreams of a similar kind about Oxford. The curious thing about these dreams is that they present two incompatible pictures of the same city, neither of which bears any resemblance to the city on the Isis/Thames. In one it appears like a spacious city of the American Middle West. The streets intersect in perfect right-angles dividing the level site into square blocks, some of which are landscaped as park-like open spaces while others contain the university buildings. This is my dream Oxford as I perceive it from the air or from a car speeding along one of its more spacious roads; but once inside it, once embarked on the task of exploring it on foot, I discover an entirely different city. Here the university buildings are huddled together more closely and irregularly than anything you'll find in Queen's Lane or the Turl. My own college of Christ Church lies in the middle of this confusion. Apart from being built of stone it has *nothing* in common with its prototype; yet any doubt that this city might be Oxford has never crossed my mind until I have woken up.

The University of Hull, where I worked for thirty-five years, increasingly figures in my dreams. Its layout is basically as it was when I first encountered it (Fig. 30). The two square buildings look much the same, but internally they are like rabbit warrens. The broad corridors are replaced by narrow passages, the stone-paved staircases by constricted near-vertical shafts with steep stairs or even ladders, giving access to higgledy-piggledy attic rooms, more like something out of Tolkien than reality.

When I leave, I usually spend time, often fruitlessly, looking for my car in one of the amorphous car-parks which bestrew the campus. One thing which never falters is my orientation. I *always* leave, whether by car, cycle or on foot, southwards, turning west into the main road, as I would in reality, for home. If I reach my house it is never in the same place but always somewhere in an unfamiliar crisscross of closely packed urban streets, and it looks quite different every time.

Sometimes real places are changed in dreams in ways which invite more obvious interpretations. Quite recently I had two non-recurring dreams which will illustrate the point. In the first I was approaching Fakenham from the direction of Wells along the Sober Road (p. 78) which was flanked on its western side by an escarpment of red sandstone several hundred feet in height. Near the top the rock itself was exposed in a prominent cliff surmounted by a large number of Scots pines, not unlike the red sandstone escarpment which I had seen many times at Hawkstone in Shropshire. The road descended more steeply into the town than I remembered it, but otherwise the place hadn't altered greatly since I lived in the neighbourhood many decades ago. The church tower, for instance, surmounted the roofs of the buildings very much as it really does.

In the second dream I was standing on the turnpike looking eastwards over the village of Stibbard, though I was under the impression that it was Diss, both being places, of course, in which I had once lived. All that was visible of the village was the church tower which rose out of a great forest. Of the fields, the hedgerows, in short the mixed farming landscape of North Norfolk, nothing was to be seen for the cover of woodland.

At my time of life I can look back on my early days in East Anglia with great affection. Memory tends to select what is most acceptable and those rose-coloured spectacles infuse it with an unreal charm. But from a landscape point of view my childhood haunts were lacking in two commodities which rank high in my list of preferences, the hills and the forest, and, although Norfolk is by no means as flat as those who don't know it are inclined to believe, I'm bound to say that the site of Fakenham was made vastly more attractive by that wooded sandstone cliff. There are parts of Norfolk also which are reasonably well-wooded, but in the north-west of the county the woodland is more sparse, as the Ordnance Survey map will show. I can only suppose, therefore, that these two dreams were, so to speak, remedying the shortcomings of places already favoured in the memory so as to make them even more idyllic.

An interesting phenomenon of the dream world as an interpreter of landscape is to be found in the area of overlap between sleeping and waking. We're not dealing here with two entirely separate images belonging to two entirely different worlds. Even in the dream counterparts of cities like Norwich and Oxford there are intermingling themes, interwoven threads. I will conclude with some examples of dreams which in different ways explore this interface between the world of reality and the world of the dream.

From childhood until early middle age I was given to the disconcerting habit of sleepwalking. I must have outgrown it twenty or thirty years ago, but while it lasted it was something I had to learn to live with. It happened chiefly in the summer when the nights were warm and days were at their longest. Even in the latitude of the English Midlands there was often still a faint glow in the northern sky at midnight.

My dormitory or 'bedroom' at Shrewsbury faced north-north-east and during one summer night I dreamt that I was looking out over the town towards the glow in the northern sky. In the foreground the detail was exactly as it should be. Some thirty feet below me or probably more there was a paved or concrete yard from which access was obtained to the kitchens, this being at the back of the School House. Beyond it a few small outbuildings and a brick wall closed it off from the steep grass slope which went down over a hundred feet to the banks of the Severn as it looped round the site of the medieval town on the farther side (Figure 19). In my dream, however, there was one remarkable feature which I had never seen before. A huge ornate staircase led down from the bedroom, passing over the wall and leading right down to the water's edge. The temptation to descend it was irresistible, but for some reason I found myself unable to move. That reason I discovered when I woke up to find that a boy in the next bed was gripping one of my ankles. The other was already outside, where the staircase had so recently been but was now no more. I soon realised that I was astride the sash window, but whether that would have been enough to restrain me I rather doubt, had not my neighbour woken up at the critical moment.

My second story of this No-man's-land between the worlds of sleeping and waking is less harrowing to look back on, but, because it occurs frequently and in a large number of different forms, it has another kind of significance. In the house I have lived in since 1960 a dressing-table with a large mirror stands in the bay window of my bedroom. A street light provides enough illumination to shine through the curtains and delineate the outline of the mirror. It's a common experience that people who wake up in the middle of the night don't realise for a moment or two where they are, particularly if they've been away for a period. I've had numerous dreams which have ended with my waking up and seeing the outline of this mirror framed by the window which the imagination quickly turns into all sorts of objects, a bus, an elephant, a boat passing under a bridge, a rock sticking out of the sea, depending on what was happening in the dream. As the real world, albeit very inadequately lit, slowly breaks through, some extraordinary confusions result!

I had one dream, back in 1965, which was of this kind, in that it involved crossing backwards and forwards between the worlds of sleeping and waking. I want to tell you about it because, although at the time I could see no particular significance in it, it later emerged as something of a milestone. Iris and I and our two younger boys, Charles and Mark, then aged eight and four respectively, spent a holiday near Tenby in Pembrokeshire, the extreme south-west corner of Wales, in the course of which we visited Pembroke Castle. While the boys played around on the castle walls I stretched myself on the grass and soon arrived at that condition between sleeping and waking in which the eye is still capable of taking in visual information, but only just. The picture communicated to the brain is rapidly pounced on by the imagination. It's neither an act of environmental perception nor a dream, more of a half-way house.

I was deeply conscious of being in a secluded, protected area, enclosed on all sides by the medieval fortifications - a continuous wall punctuated here and there by gateways, turrets and the occasional flight of steps. Needless to say the boys were quickly on the top of the wall, and as I watched them they dissolved into the silhouettes of medieval soldiers. They were the watch, a kind of extension of my own faculties of perception. The top of the wall was the horizon of *my* world. *They* could see beyond into the wider world outside; and the thought crossed my mind that, for hundreds of years, people within this sanctuary must have felt a similar sensation, though it's not likely they would have verbalised it in quite the same way.

Since I have a propensity for falling asleep at the drop of an eyelid, so to speak, and since it was a warm summer afternoon, it wasn't long before I found myself in the middle of a real dream. I was back in a place which I had first visited with Iris in 1951. We had stayed for a few days at a village called Arzl-im-Pitztal in the Tirol, and in my dream I was lying in a meadow beside the little Pitz Bach which gives the valley its name. Beyond the immediate foreground the pine forest provided a screen of dark shadows, and beyond that the wooded valley sides swept upwards, broken only by the alp where the gradient eases and from which the trees had long ago been cleared away to make the high-level summer pastures. Beyond the alp the forest continued for a few hundred feet before giving way to the bare rock on the skyline.

But somehow I wasn't there at all. I was still lying on the lawn encircled by the walls of Pembroke Castle, and for a moment I was equally in both places at once. The rim of the distant mountains and the top of the castle wall had become one and the same thing. What persisted as a continuing reality linking the Austrian valley with the Welsh castle was the feeling of being down there in a quiet, cosy place, and knowing that the sense of security came from being surrounded by an encompassing, enclosing rampart. It was like being in two theatres of vastly different scale, watching the same play at the same time and getting the same message. 'Pembroke-im-Pitztal' went into the storehouse of remembered places and was to become quite important in my later attempts to rationalise the aesthetics of landscape, as we shall see in Chapter 9.

My last example involves the carrying-over of a problem from the real world into the world of dreams in search of a solution, and is unusual in that it was followed by my actually doing something about it. I painted a picture. I've already described the problem (p. 69). The view to the north from the windows of the School Building at Shrewsbury was compulsively fascinating, not least when compared with Cicero or Thucydides. It also acquired bonus points for being out of bounds and therefore more mysterious. The greater part of outdoor school life, however, was lived on the side facing south-south-west (Figure 19), and it was only from this side that the main building could be entered. The whole of this area, The Site, was cut off by buildings and/or trees from the view over the North Shropshire Plain. It was like living on a table with an uninterrupted line of large objects placed contiguously along the edge. One was always conscious of that spectacular drop to the Severn and of the distant view over the town and into the distance, but could never actually see it.

In my dream (Figure 39) I saw the front of the School Building as it appeared from The Site, clearly recognisable yet altered in four respects. First, all the sash windows had been removed so that the interior of the building became a cavity of darkness, a non-place. Secondly, the level land in front of the building had been raised so as to provide a vantage-point high enough to afford a view over the roof. Thirdly, the chapel and trees on the left-hand side had been removed to allow a view past the side of the building. Fourthly, and most dramatically, a great arched cleft had been opened up in the middle like a huge keyhole, a sort of eyelet permitting a peep through the very centre of the building. The irony was, however, that the view revealed by these structural changes had itself been eliminated. Not only the town but the whole North Shropshire Plain had been replaced by a featureless expanse of emptiness. Only vestigial suggestions of distant escarpments remained from the world of reality. Was the subconscious telling me that I didn't really *want* those two pictures to be integrated and the enduring mystery to be finally resolved?

6   FACT AND FANTASY

'Imagination is that sacred power.'

This chapter will carry our investigation of individual variation a stage further. Those fragments of the real world which, as we have seen, have already been greatly altered before they become available as our building materials, have to be cemented together by imaginations which are no less variable as between one person and another. I shall take three extended examples to show how this has applied in my own world-making.

The first example concerns an inordinate, almost obsessive interest in a particular subject. This is anything but unusual. Such interests are found in everybody though their intensity may vary greatly. Hobby interests like fishing, sailing, rock-climbing, ornithology, even stamp-collecting have the power to invest particular places, or kinds of places, with special values which would only be shared by others with similar tastes. In my case the obsession was with railways. My second example will deal with fantasies which are rejected rationally but still accepted emotionally, and in this context we shall have a further look at animism. Finally we shall see what happens when landscape images spill over, as it were, into some other 'world', and I shall illustrate this by reference to music.

What I mean by an obsession is an interest which leads people to spend so much time, effort, energy and often money in the pursuit of what would generally be regarded as an unimportant objective as to suggest that their sense of perspective has been over-ruled by some fascination of an irrational and probably emotional kind. I've been subject to a good many milder interests of this kind, but, apart from 'the landscape thing', only one which could merit the description of an obsession.

Railway Mania

I've already said that, in my case, the interest in railways didn't find expression in the making of models, or, if it did, this was only to a very limited extent. Neither was I ever interested in collecting the numbers of engines, a hobby which must have consumed millions of leisure hours, not only of boys but of grown men and occasionally even of women. This is probably because my collecting instinct is as weak as my sense of number. I distinguish between a collecting instinct and a hoarding instinct. The former is a positive, purposeful, orderly habit, involving some measure of classifying, marshalling and arranging. The latter is its negative counterpart, born partly of indolence, partly of a disinclination to make irreversible decisions and partly of a misplaced sense of thrift, not to mention a gut feeling that the irresponsible act of throwing something away will be punished by the early discovery that it was exactly what one most needed.

For me, therefore, number collecting was even less viable as a hobby than model making. I can at least become excited by the models made by others, and, had I shown a less parsimonious attitude towards the cost, I might easily have become gripped by the challenge and have emulated some of my friends in filling their attics with new-created worlds of points and crossings, tunnels and viaducts and a kind of infill of cardboard shops, trees and miniature public open spaces. In the event I chose my own brain as a preferred alternative to my attic, and it was there that I began the construction of a complex, and to me, very exciting model.

When I first began to notice railways the route-system of Great Britain, which had been developed over the previous hundred years by a large number of competing railway companies, had recently (1923) been amalgamated into four large groups ('The Big Four'). It took a good many years, however, for all vestiges of the previous pattern of ownership to disappear, particularly with regard to locomotives and rolling stock. These, even after they had lost their old distinctive liveries, retained unmistakably the design features proper to their places of origin. The larger of the old companies, and even some of the smaller ones, had employed their own mechanical engineers with their own staffs to design and manufacture their own fleets of locomotives suitable for handling different kinds of traffic over their own route-systems. The necessity for pulling trains of different weights at different speeds over different distances along lines characterised by different gradients and different degrees of curvature, and all subject to different financial constraints, had resulted in the Big Four companies inheriting a large number of locomotives of many different classes.

Often the only mark of differentiation by which the layman could distinguish between locomotives was the unique serial number, except for the larger, more prestigious, more aristocratic engines which were dignified with nameplates; but to the trained eye the classes themselves could be readily distinguished by their shapes. Particular engineers had a predilection for particular design features, so that it was common and meaningful within the fraternity to speak of, say, 'Horwich lines', meaning a delineation (what I suppose the animal breeders would call a 'conformation') characteristic of the practice of successive engineers of the former Lancashire & Yorkshire Railway Company who built their engines at the works in Horwich, not very far, as it happens, from Bradley Hall.

You have presumably realised that I'm speaking of steam locomotives. By the early nineteen twenties electric trains were well established but only in certain parts of Britain, especially the South-east, and only on an occasional visit to London did I encounter them. When I did I regarded them with interest but never with passion.

The steam locomotive, on the other hand, was capable of lifting me to emotional heights which few other stimuli could rival. I suppose that this had something to do with an early acquaintance with railways in the pleasurable ambience of Bradley Hall and with the rose-coloured spectacles of nostalgia. It had something also to do with an awakening understanding of simple mechanical processes which were visibly demonstrated in cylinders, pistons, cross-heads, connecting-rods, coupling-rods, valve gears and the other paraphernalia by which the unseen power of heat-energy was transmitted through the reciprocating motion of pistons to the rotary motion of driving-wheels.

I say 'unseen power', and this is strictly true, because heat-energy in superheated steam is invisible; but the manifestations of that power were dramatically expressed, not only in the rhythm of the moving parts, but also in the pulsating exhaust-beats which blasted steam and smoke high into the sky, each beat accompanied by a kind of gasp of exhalation, bringing the infant ear into an exhilarating partnership with the infant eye. Even quite a young child, initiated into the mysteries by a well-informed parent or uncle, could tell a three-cylinder from a two- or four-cylinder engine. It puffed in six-eight time. How, I wondered, could anyone be so obtuse as to suppose that a steam locomotive is not a living organism?

You won't be surprised, in view of my earlier references to animism, that the most potent ingredient of the locomotive experience was precisely this. All locomotives are essentially vehicles; they differ enormously in their capacity to become something more. A graceful ship can achieve a kind of personality, even a sexual identity, always distinguished by the feminine 'she'.

An aeroplane can, with a minimum of help from the imagination, become a bird, because it employs quite ostentatiously the same aerodynamic devices, wings, a tail and a streamlined configuration to enable it to overcome the force of gravity. By contrast an electric or diesel locomotive is destined to remain a mere vehicle. The means by which it develops and transmits its energy are concealed within a largely opaque rectangular box. Often it can develop greater power than a steam locomotive of comparable size and move at greater speed, but it can't lay hold on the gift of life!

The concept of a locomotive apparently endowed with a capacity to move of its own volition, to accelerate or to slow down on the recognition of a given signal, to control as if by force of personality the behaviour of the less potent vehicles committed to its charge (so eloquently described by the Reverend W. Awdry in *Thomas the Tank Engine* and the other stories in his immensely popular series) invites a kind of ecological extension of the animistic image. Most animals can consistently be associated with their own proper habitat, a type of environment, often restricted to definable geographical limits, to which they're naturally attuned and within which they're habitually confined, not by constricting barriers but by inclinations, characteristic of their species, to frequent places conducive to their respective life-styles.

Each class of locomotive had its own proper territorial range, so clearly defined that it could be precisely delineated on a map. It's true that the companies newly constituted in 1923, which had inherited fleets of vehicles from their constituent companies, found it convenient to use them on certain stretches of track which under the old order had belonged to alien managements. Some Great Eastern engines, for instance, built for East Anglian lines, now regularly worked in the North of Scotland, but most of these adjustments had already been made before I understood what was happening, so they seemed natural enough. Only rarely did a locomotive stray on to the rails of another of the 'Big Four' companies except when exercising running powers under long-established authority. Occasionally exchanges were arranged to discover whether locomotives could operate efficiently on other companies' lines - a Great Western engine, for example, on the L.N.E.R. line from London to York, Newcastle and Edinburgh - but however successful the experiment from a technical point of view, it looked radically wrong. A Bengal tiger strolling down Oxford Street would scarcely have looked more out of place.

During these formative years I rarely travelled on a train myself; the money didn't run to it. Now and again the need to perform some necessary journey justified the experience at no extra cost. Three times a year from the age of thirteen I crossed the bosom of England, there and back, on my way to or from Shrewsbury School, but quite often my mother availed herself of the excuse to visit my Uncle Jim and Aunt Elsie in Herefordshire, and we travelled by car. So my developing picture of the railway on the ground was pieced together from views of the railway rather than views from the train, and this of course is the right way to do it. In those days with few exceptions there were no diesel multiple-units or other vehicles which allowed the passenger to see backwards or forwards along the track. I was nearly forty before such an opportunity came my way. In short, the windows of the railway carriage commanded a view of everything in the landscape *except* the railway.

The only way to see along the line from a moving train, except on a sharp curve, was to put one's head out of the window, a dangerous practice against which passengers were invariably warned by a conspicuous notice. Notwithstanding this, in my late 'teens, when I should have known better, I once donned a pair of motor-cycle goggles and travelled the 179 miles from Euston Station in London to Chester with my head out of the window on one side of the train or the other all the way, apart from a few interruptions when the train was in a tunnel. As we pulled into Chester General Station I slipped into the loo to tidy myself up in anticipation of meeting Mrs Gough, my hostess, mother of my schoolfriend, Hugh, with whom I was going to stay. There I removed the goggles and looked in the mirror, to be greeted by a plausible imitation of a black-and-white minstrel. Such were the lengths to which I would go to perfect my knowledge of the West Coast Main Line.

Generally speaking, however, my experience of views *along* the line was confined to places where railways were crossed by roads or footpaths, at bridges or level crossings, or in stations. From such vantage-points the metal rails disappeared into unattainable territory, sometimes round a curve in the foreground, sometimes straight as an arrow shot into the far distance, but always, eventually, out of sight. The railway, in short, symbolised the means of going from where one was to somewhere else, somewhere unattainable except by the train itself. The distant vanishing-point of a pair of apparently converging railways lines is an intrinsically romantic thing.

During my 'teens, when we made a number of journeys to most parts of Britain, I explored its railway network by car. My long-suffering parents and sister were frequently expected to defer a picnic for an hour or two so that we could reach some idyllic spot, pre-selected from the map, which commanded a good view of a busy main line. There we would eat our sandwiches seasoned with the sulphurous tang of railway smoke. For me the eating of sandwiches was a secondary occupation which had to be fitted in with watching for the on-coming trains, trying to recognise at the earliest possible moment the outlines of their engines and to attribute them to their proper classes, counting the carriages, and, if they weren't moving too quickly, reading the headboards proclaiming their destinations, or, if lucky, announcing their titles, such as *The Queen of Scots Pullman, The Irish Mail, The Cornish Riviera Limited* or whatever.

In this way I gradually pieced together a picture of the British railway network, and it was this rather than the engine numbers or the precision scale models which mesmerised me. I was particularly intrigued by three aspects of it, three different ways in which it could be perceived, and you must either bear with me while I amplify them or skip the next few pages.

First, the railway network was a web, each thread of which suggested a link with some distant place. As I became better acquainted with the system, so the various destinations came to mean more. Places whose relative positions had become familiar when I had visited them by road were also linked by an independent alternative system. The more one's knowledge grew, the deeper became the symbolic meaning attached to whatever point in the network one happened to encounter.

The passage of a train along such a thread brought this sense of interconnection even more vividly to life in much the same way as the playing of a piece of a familiar tune gives a moment of heightened reality to something which, in a sense, exists all the time. There would be a strong sense of anticipation as a train approached from familiar origins in one direction and of vicarious adventure as it headed for familiar destinations in the other. If origins and destinations were not familiar from first-hand experience this added an acceptable sense of mystery. Places which up to that time I had never visited, like Grimsby, Aberdeen and Bognor Regis, were represented in the mind's eye, so to speak, by substitute images, visual impressions built up, sometimes quite erroneously, from postcards or other pictorial evidence and, of course, the study of the Ordnance Survey map.

I'll give you one example of such an experience, though I lack the skill to convey in words the intensity of the exhilaration which it produced. I was camping in Glen Garry near the Falls of Bruar in Perthshire, Scotland. Rising at about the same time as the August sun I took a short walk through the dewy meadow to the railway about half an hour before the northbound *Royal Highlander* was due. Having taken up a position beside the line I waited for its arrival in a state of eager expectation. The line at this point runs more or less from east to west, so the light easterly breeze was coming from the direction of Blair Atholl, and my first intimation of the advent of the night sleeper from London was the sharp puffing of the engine as it drew out of the station some four miles away down the valley.

The admission of steam to the cylinders of a locomotive is controlled by the regulator, which is to all intents and purposes a tap in the main steampipe, and by the reversing-gear which is also used to determine the point in the stroke of the piston at which steam is cut off by the valves. When the engine starts, the cut-off is opened up to its maximum (usually about sixty-five percent of the total travelling-distance of the piston), while the regulator is opened only a little, because the full, unrestricted pressure of the steam on the pistons would otherwise cause the wheels to slip and spin. If you're old enough to remember you will no doubt call to mind how, every now and then, they would do just this to the accompaniment of a great roar of rapidly accelerating exhaust-beats which the driver would check as soon as possible by closing the regulator and trying again. Once the train is on the move, however, the economical use of steam requires that the cut-off be made earlier, thereby restricting the admission of steam to the first part of the stroke and allowing its expansion to continue to work during the latter part. Only when this adjustment to the cut-off has been made is it practicable to open the regulator fully without fear of wheel-slipping.

So it was on this morning that, alerted by the first few puffs, I soon lost the sound as the cut-off was notched back, and two or three minutes elapsed before I picked it up again. By now it was much nearer and the rate had quickened, though the individual puffs were clearly quite distinguishable because the steepness of the climb kept the speed down to about thirty miles an hour. The line ahead was rising to some fifteen hundred feet above sea-level at Drumochter, the highest main-line summit in the British Isles, so the engine was making a great deal of noise which reverberated through the glen. Eventually a plume of white appeared above the trees, and in a couple of minutes the train was in sight.

The climax of the experience came at the moment when the train was passing only a few feet away. The procession of gleaming carriages in deep Midland red was pulled by a shiny black engine which was snorting up the gradient and filling the sky with huge spurts of steam that rapidly condensed into white, crumpled clouds in the damp morning air. The intensity of what would in any case have been an exciting moment was increased by the instantaneous release of a great store of remembered images, pigeon-holed in my memory over the years as I had acquired a detailed knowledge of the route followed by the train. It had left Euston Station, a familiar and favourite haunt of mine, the previous evening bound for Inverness, and in the night it had passed innumerable places which I had come to know like old friends, including the reservoirs at Marsworth, where I had walked with my Auntie Amy, and the series of picnic sites where the line was crossed by alternative roads between

Stibbard and the Welsh Borderlands. Somewhere about midnight it
had passed the rough meadow which separated the railway from my
dear little garden gate at Bradley Hall. All these images and countless
more seemed to come together in a single momentary flash of
apprehension, as if bits of them had been brushed off and collected
up by the passing train which was now offering me the chance to
grab them and feed them into my own system.

I'm well aware how difficult it must be for anyone who has not
enjoyed the same combination of experiences to understand how a
sense of geographical unity coupled with a romantic passion for
steam engines can be a potent enough idea to trigger off such a deep
emotional sensation. The point I am trying to make is this. The more
peculiar the whims and fancies of one lone youth on a Scottish
fellside at sunrise may appear, the most unreasonable does it become
to suppose that we can offer universal generalisations to explain the
ways in which we collectively perceive the world around us without
making allowances for the highly subjective nature of our own
individual modes of perception.

The second way of looking at railways which fascinated me
was to see them as the products of their own evolutionary histories.
This phase came a little later, and I must have been nearly twenty
before I reached the peak of my enthusiasm for reading railway
histories. Once I had started, however, I devoured every company
history I could lay hands on. By the time I had completed my
curtailed course in geography at Oxford my particular delight was
to discover the various processes and stages by which the network,
now fairly familiar to me, had grown. Nearly all histories of
railways at that time dwelt heavily on the personalities who had
shaped them, the financial and political backgrounds to their
circumstances of origin, the struggles and squabbles between their
rival managements, the intrigues, liaisons and plain chicanery, in
Parliament, in the boardrooms, in the country houses and in the
London clubs, often gave little indication of where these
controversial lines were actually supposed to go! Maps, if included
at all, were generally unreliable and invariably inadequate;
sometimes they were non-existent, so I learnt to read the historical
record hand-in-hand with the appropriate Ordnance Survey maps.
This, of course, is the proper way to do it, because the map
furnishes an immense amount of additional information and
frequently suggests an explanation of some paradoxical question
which had entirely escaped the author; but one does need enough
geographical information from the text to make possible
identification of places on the map without ambiguity, and even the
most reputable railway histories generally failed to some extent in
this respect.

FIGURE 40. 'TWO SETS OF VARIABLES'. The West Highland Line between Fort William and Mallaig has to find the best possible accommodation with a highly irregular land surface shaped by natural forces operating over millions of years. Photo by the author, 1963.

This marriage of text and map brings me to the third 'way of looking'. My studies in geography at Oxford had been largely directed towards the physical side, and, although this emphasis was to change, I never lost my initial interest in the processes which had shaped the configuration of the land. I became intrigued by the challenge of bringing together two sets of variables, each controlled by a very different set of principles (Figure 40). On the one hand there were the physical components of the landscape, the hills and valleys, the river terraces and the intervening bodies of water which either facilitated the passage of the railway or imposed in its path obstacles of varying severity. The processes which produced these landforms were the province of study of the geomorphologists. On the other hand there was the network itself, composed of constituent parts which had come together through historical circumstance and whose capacity to deviate from the straight line and the horizontal plane was constrained by a number of definable technical limitations. It was this area of research which dominated my academic work for the first twenty years of my career as a university teacher and I shall be referring to it again in Chapter 8.

I suppose the satisfaction to be found in most hobbies will be likely to satisfy more than one kind of desire or need. This obsession with the transport system, which has claimed thousands of hours of my time, seems to bear this out. On the one hand the fitting together of all the pieces of the pattern, which clearly gratifies my need for orientation, leads to a satisfaction which certainly has an intellectual dimension. The pleasure is a rational pleasure, the pleasure of knowing, of understanding, of having found out. On the other hand the animistic image of a steam engine is anything but rational. It belongs to the world of fantasy, of the imagination unconstrained by those checks and balances by which the reason seeks to keep it in bounds. Since, after my appointment at Hull, this obsessive hobby became the basis of my academic research, either the emotional and the irrational aspects had to be banished for ever or I should have to find a way of keeping them strictly, if only temporarily, out of the way while engaged in research. The former would have involved a complete personality change which I would surely have found impossible to contrive, so it was fortunate that I managed, at least to my own satisfaction, to achieve the latter. Fantasy, in any case, went much deeper and spread much wider than the railway mania; so let's turn to the second of these areas in which it impinged so strongly on my involvement with the landscape.

The Phoenix of Animism

The strong streak of animism which lay so thinly disguised just beneath the surface of my obsession with railways was, I suspect, simply a development of a very widespread, in fact universal phenomenon, though the form which it takes must be peculiar to each individual. All children in all cultures are encouraged by nursery folklore to blur the distinction between animate and inanimate objects. In what we patronisingly call 'primitive societies' animistic conceits persist into adult life as is testified by whole libraries of anthropological literature. Where we are in danger of making a serious mistake is in supposing that either individuals or societies completely grow out of the animistic stage. As soon as we think we have killed it off by the superiority of our powers of reasoning we find it rising again like the Phoenix from the ashes of its own destruction.

Adult members of 'advanced' societies may no longer believe that the making of mountains and other features of physical geography is the work of a giant or of the devil, but they will conceptually minimise the difference between people and things in all sorts of ways. A motorist will see his car as an extension of his own person and the controls almost as a part of his own neural system. He will say such things as 'I'm too wide to get through' or 'I thought he was going to hit me'. In both cases he will be referring to his car and not his person, but the use of the personal pronoun implies that he is speaking of them as if they were one and the same. A golfer will use the first person when referring not to himself but to his ball. 'I think I'm on the green' he'll say, when he knows perfectly well he's two hundred yards away. All this may seem a long way from animism in the mythological sense, but it shows how easy it is to allow animistic imagery to penetrate our habits of thought and speech without ever realising what's happening.

I've already made a number of references to some of my early experiences of animism. The two trees on the Stibbard turnpike, you may remember, when viewed from my bedroom window, coalesced into that friendly beast which was already munching its way along the hedge before the dawn of my memory. But other trees there were which, to varying degrees, assumed some kind of personality. The handful of conifers of exceptional height whose spires surmounted the canopy of the Fulmodestone Severals seemed to be a sort of aristocracy, a ruling class. It was no difficult matter to ascribe to them qualities like arrogance and attitudes like superiority towards the lowlier members of their own tribe on which they looked down with disdain, while at the same time inspiring a sense of confidence in their capacity to use their advantageous position, like look-outs, for the benefit of these same inferior neighbours.

The various trees in the garden were different in the same sort of way that people were different. The beech trees which separated the rectory from the church tower were aloof and stand-offish. The fact is, I suppose, that I couldn't manage to climb their smooth trunks and therefore never felt the intimacy of the sheltered spaces within their foliage and their enfolding branches, unlike the oak tree and the chestnuts on the front lawn which therefore appeared more friendly. The old ruined stump which overleaned the road at the bottom of the garden simply oozed friendliness. The rotted cavity of its great bowl could hardly have invited us more eloquently if it had been possessed of the power of speech.

Trees, after all, are endowed with life, even if it seems to be quite a different kind of life from that enjoyed by the members of the animal kingdom. But mountains? At first sight it may appear that hills and mountains are even more improbable as repositories of souls, yet they have many qualities which render them suitable candidates for the attribution of personality. It's true there are huge scale differences, but this isn't in itself important. Tales of giants and fairies, of mice conversing with elephants, of djinns, newly released from bottles, towering into the sky like massive cumulus clouds and talking down majestically to earthbound mortals, all these had paved the way for an acceptance of the fact that enormity was not in itself an obstacle to intercourse, though it did follow that, if mountains were to be perceived as personalities, they could never be approached quite so intimately as, say, the yew hedge in which I daily experienced the sensation of being totally enveloped.

Sometimes the outline of a hill presents quite graphically an animate image. I suppose the Lion and the Lamb at Grasmere was the first hill I ever saw which displayed an unmistakable likeness to a living creature, or in this case two living creatures, if only in silhouette. But the resemblance to an actual animal didn't have to go so far in order to evoke the feeling that the mountain was alive. Provided that its peak was sufficiently detached from its neighbours to establish its individuality as a separate entity it could qualify. Probably it would even have a name. Most of the hills encountered in my first visits to the Lake District were introduced to me by name. My father, who had walked them many times, knew them all, so I was confronted with the name and the image at one and the same time, just as when I was introduced to the neighbouring clergy. Like people, too, they looked different when seen from different sides. Not only that but they changed their moods so dramatically. When the sunshine gave way to the cloud shadow the whole mountain seemed to be expressing another side of its personality. Sometimes, too, it would bury its head in the cloud and then it would assume an air of mystery like the king in the story-book who is known to be in the castle but who can't actually be seen. Speculation, conjecture and wonder endow the unseen summit with a certain awe, the memory of which becomes built into the image and remains part of it even when it appears on another occasion crystal clear in the sunshine.

When I went to Shrewsbury at the age of thirteen I had certainly not outgrown this way of looking at the hills. If anyone had taxed me with the allegation that I was practising animism I should probably have strenuously denied it, that is assuming that I knew what it meant. But as I became acquainted with the various hills of the Welsh Borderlands I found myself developing quite different attitudes to each of them individually, different feelings about them, different emotional responses. They tended to possess strongly contrasting features, so that the attachment of individual personalities followed more easily than if they had all looked similar.

The Wrekin (Figure 41A), which lies about ten miles east of Shrewsbury, was probably the most conspicuous. Thanks to its isolation, and therefore the absence of any intervening obstacle, it can be seen from many miles away. It had attracted colourful mythological stories about its origin involving a Welsh giant; it exuded a kind of charisma which in turn secured for it a symbolic status within the surrounding region. 'Friends around the Wrekin' was a sort of slogan, a rallying-cry to stimulate local patriotism and furnish the excuse for a toast, no doubt in Wrekin Ales. As seen from Shrewsbury its profile rose gently from the left and plunged more steeply to the right where lay the Severn Gorge, out of sight but never out of mind. For me a particular significance attached to its situation; it lay between me and the main body of Midland England, between me and Stibbard, if it comes to that. It seemed constantly to be saying

> Are you really comfortable over there? Are you sure you're not in Wales? If ever you feel lonely take comfort from the fact that I can see Wolverhampton from up here, Stafford even, and that means the L.M.S. main line from London to Bradley and beyond, and you couldn't have a more intimate friend than that, could you?

The fact that the Wrekin lay north-east of the Much Wenlock road, and therefore out of bounds, probably helped. People tend to provoke admiration, at least among schoolboys, when legitimately occupying places where they are not allowed to go themselves. Perhaps it's the same with mountains.

To the south of the Wrekin, some dozen miles away, lay the Clee Hills, first the Brown Clee and beyond it the Titterstone. They were

FIGURE 41. SOME PERSONALITIES OF THE WELSH BORDERLAND. A. The Wrekin from Lyth Hill. View to the east. B. Caer Caradoc from the north-east. C. The Stiperstones. Part of the 'dorsal fin' of the steeply inclined Stiperstones Quartzite. View to the north-west. D. Pontesford Hill from the north-west. This has undergone a complete personality change since the nineteen thirties, thanks to the removal of its most conspicuous feature, the wispy 'mane' of conifers, and the extensive new planting on its flanks. E. Corndon Hill from Pennerley. View to the south-west. F. The Breidden, surmounted by the Rodney Column, slopes steeply down on the right to the River Severn. View to the south-west from Melverley Green. Photos by the author, 1993.

41 B

41 C

41 D

41 E

41 F

both rather higher than the Wrekin and, like it, they appeared as bastions, eastern outliers of the Welsh Marches. This was how Housman had seen them as perceived from the other side when he described the feelings of the Shropshire lad, destined to go and seek his livelihood in London:

> As through the wild green hills of Wyre
> The train ran, changing sky and shire,
> And far behind, a fading crest,
> Low in the forsaken west
> Sank the high-reared head of Clee . . .
>   (Housman, 1896, XXXVII).

Similar as were their situations and their symbolic functions as sentinels keeping watch over the English Plain, the Clees were quite different from the Wrekin in personality. They were somewhat more rounded and in an indefinable way somewhat less involved. The phrase 'Friends around the Clees' would never have had an authentic ring. They lay there like a couple of fat cats perpetually sleeping off a prodigious feed; but I dare say what made them really special was a nostalgic link with a much earlier period of history; not the Roman Occupation, not the Middle Ages, but that time long, long ago when I went to stay with my Auntie Margaret in Abdon.

Caer Caradoc (Figure 41B) was quite different again. Here it was the Roman Occupation which made it so. I knew before I ever saw it that it boasted a Romano-British camp on its summit and indeed took its name from the leader of an ancient British tribe, latinized into Caractacus. Even from a distance of some miles it was possible to distinguish the irregularities of its outline where the earthworks ringed its summit. In Caer Caradoc the historical association seemed to take precedence over all its other characteristics. It had a cover of rough grass, a good view and a unique geographical position, just like all the other hills of South Shropshire; but it had something more, a personal link, almost an identity, with a historical character. One almost wondered whether those circular ridges near the summit were the marks of chafing when it had been taken away to Rome in chains.

Its immediate neighbour to the west, the Long Mynd, was, as its name implies, more elongated; it was also marginally higher, but for me it never had quite the same personality, which is something quite different from its attractiveness. The Carding Mill and other valleys were quite dramatic, and the views from the top as fine as any in the Welsh Borderlands, but as a hill it was too amorphous to establish its individuality. It even had a road going over the top. Further north-west the similarly named Long Mountain had even less personality.

Its flanks were too much encumbered with hedgerows, fine enough in the English lowlands, but not the stuff of which mountain landscapes are made. Between the two lay the Stiperstones which had more character than either. It derived its distinction as well as its name from the little outcrops of quartzite (Figure 41C) which projected from its long curvaceous back and gave it the appearance of a huge fish with a line of dorsal fins, and it therefore had less difficulty in projecting an animistic image. The same would go for the much smaller but very conspicuous Pontesford Hill (Figure 41D), which in those days carried a thin, wispy row of conifers right up to the skyline from its northern end to the summit. I suppose one could have seen them as a single line of marching soldiers, but to me they read like a mane or crest surmounting a couchant beast.

There were two other hills which possessed strong though very different personalities. They shared the common function of sentinels overlooking the Vale of Powys, as that part of the Severn Valley is called, much as the Clee Hills kept a watchful eye over its lower reaches below the Gorge, and they were both just in Wales, though only because of the confusingly deviant behaviour of the boundary at this point. The first was Corndon, (Figure 41E) a shapely if shallow cone standing somewhat aloof. Because of its isolation it commanded perhaps the best view of all over the mountains of North and Central Wales. The other was the most characterful of these hills of the Welsh Borderlands. The Breidden, or 'Breiddens', as it was often called, because it consisted of three peaks, was really spectacular on the north-western side where it plunged almost vertically into the Severn (Figure 41F).

When, some years later, I was introduced to the mysteries of geology, I came to understand that some of these personality differences were attributable to strongly contrasting geological structures. I learnt also that some of the earliest attempts ever made anywhere in the world to work out the geological succession had been made among these very hills. I won't weary you with further technical details, neither will I enumerate the many other hills and mountains which came to acquire personalities within my system of things. If I've sketched the characters of this little group it's because they were a group, highly diverse personalities sharing common membership of a family - just like people. The hills of the Lake District comprised another such family, and when I eventually reached the North of Scotland I found some dramatic examples of mountains of really striking individuality, like Suilven, Canisp, Ben Hope and Ben Loyal.

I must mention just one other Scottish mountain further south because it always held a special fascination for me. From certain directions Ben Cruachan also resembles a huge cat lying between

Loch Etive and Loch Awe, gazing out over the surrounding hills. I've climbed it only once and that was in the middle of the night, the objective being to see the sunrise from the summit. My companion was a school-friend, Tony Chenevix-Trench, who was staying on the Isle of Mull. He it was who, having gone up to Christ Church a year before me, was chiefly instrumental in persuading me to go to that college. He used to tell me he couldn't make up his mind whether to become Prime Minister or Archbishop of Canterbury and regretted that it was not realistic to suppose he could combine the offices. In the end he contented himself with becoming Headmaster of Eton, but that lay well in the future on the other side of a Japanese prisoner-of-war camp. The anticipation of seeing the sunrise anywhere in Scotland is usually an act of hope rather than of faith, and our immediate objective was only imperfectly attained; but the experience of climbing through the darkness was one which left an indelible mark, and if the image of Ben Cruachan still has animistic overtones, they are those of an essentially nocturnal creature.

I've made several other expeditions to see the sunrise from mountain tops, the most successful being from Snowdon, but only at the third attempt. The first two ended in shrouds of mist and we could see no more than the grey wet rock at our feet and the spartan stalks of fescue grass shivering with cold as much as, perhaps, ten yards away, but the third time proved proverbially lucky. Mountains in the night seem to reveal aspects of their personalities never disclosed to those who see them only in the daylight hours.

Another type of geographical feature which lent itself to a similar kind of animistic interpretation was the island, particularly if it was small enough to appear as a unit, an entity clearly recognisable as an island. The designation of such a unitary body by its proper name confirmed its individuality; where it had to share itself between two or more descriptive names it suffered from an identity crisis. I've never been to that island in the Outer Hebrides which is split by a land boundary into Lewis and Harris, but from the map evidence it always seemed to be a schizophrenic island if not a Siamese twin. Uncomplicated islands like Skye or Mull have a better chance of establishing personalities. Better still if they are smaller and sport jolly little names like Rum, Eigg and Muck. Best of all they're just stacks, like the Bass Rock, a lithograph of which hung in our dining-room at Stibbard all through my childhood.

But it's those little stacks suggestive of the erect human form which most inevitably attract human descriptive names, like the Parson and Clerk, Old Harry and his Wife or the Old Man of Hoy. Almost everybody can relate to these on familiar personal terms, but by no means everybody can generate the same sort of feelings about those more ill-defined, amorphous regional entities, unendowed with any anthropomorphic advantages, like the Fens, the South Lancashire Coalfield or the Basin of Carlisle.

Rivers came into a special category of inanimate objects into which my little imagination breathed the breath of life, but in this I could perhaps be said to be almost normal. The attribution to rivers of not only human but even divine powers is timeless and world-wide. Old Man River, Old Father Thames and Mother Ganges come to mind together with innumerable lesser streams and rivulets which have not only housed but even been water gods or goddesses. As candidates for animation rivers have the great advantage of consisting of a substance which is constantly on the move, and movement is powerfully suggestive of life. There is, of course, room for confusion between the river itself, the river-bed and the river-valley, but these concepts, though quite distinct, are so closely related that this doesn't seem to matter, and the same name is just as likely to evoke the image of swirling water as of rock ledges, willowed banks or curving valley sides leading the eye as well as the stream out of the field of vision and into the romantic faraway.

There were other animistic undercurrents of all sorts to be found not far beneath the surface of my reconstruction of the world. Clouds and sky-images, for example, were a fruitful source; but if this strikes you as proof of extreme infantilism, eccentricity or even mental derangement, reflect for a moment how difficult it is to draw hard-and-fast lines of demarcation between the literal and the metaphorical usage of words and phrases as we commonly employ them. Such statements as 'The Stock Exchange was nervous', 'Wall Street was in an optimistic mood' or 'The Quai d'Orsay was furious' are all metaphorical statements attributing human feelings to inanimate objects. What we mean is that certain people within these places or institutions have acted in the way we have described, but what we say is that the places themselves have done so.

Again, whenever we make an inanimate object the subject of a verb of motion we invite animistic metaphors. If we say, for instance, that a road climbs a hill we're using an expression which is so common that we're unlikely to be aware that it is a metaphor at all; yet what we mean is that people or vehicles climb the hill using the road as a means of passage. The road itself, being fixed to the ground beneath it, can never do so. When we stop to think about it we soon realise that all societies, even the most advanced and sophisticated, are up to their necks in animistic symbolism. How's that for an animistic metaphor, by the way?

There are three points arising out of this brief allusion to animism which I shall ask you to remember when reading the following pages. The first is that my mental picture of the world around me has been invaded by innumerable animistic images of all kinds; I've only disclosed a fraction. The second is that it's so difficult to draw clearly defined boundaries between what is metaphorical and what isn't that any attempt to deal with this problem by attributing particular cases to particular categories is likely to prove extremely dubious if not meaningless. The third is that the animistic imagery which has entered into the painting of my mental pictures is unlikely to resemble, except in the broadest outline, the animistic imagery which has entered into yours, but that doesn't mean you're not up to your neck in it too!

Melodic landscapes

The third and last of my examples under the general heading of 'fantasy' proceeds from the fact that we may find ourselves constructing not just one world of our own making but many, and that, as they take shape, one such 'world' may influence the development of another. The relationship between landscape and music is a case in point.

By no stretch of the imagination could I be said to come from a musical family. Both my parents enjoyed what might be called popular light music, but their knowledge of even that was limited. My father would also have claimed to know a certain amount of religious music, some of the choruses and arias from *Messiah* and a handful of other well-known numbers from sacred oratorios and a very few church anthems, but on the whole he found these unconvincing partly because they kept on repeating the words, which, he maintained, was both unnatural and unnecessary! As a clergyman, of course, he knew plenty of hymns and chants. Military bands were all right because they invariably played something which went with a swing. Opera was out!

Well, nearly out! There was one exception to the general rule that my father was not really interested in music. He had, like many other non-musicians of his generation, a deep affection for, and a considerable knowledge of, the works of Gilbert and Sullivan. He would sing or whistle airs and choruses from all the operas on every kind of occasion, but when as a very small child I was carried up to bed on his shoulders it was always to the accompaniment of a favourite from *The Gondoliers:*

> In enterprise of martial kind
> When there was any fighting,
> He led his regiment from behind
> And found it less exciting . . .

Every night he would wade through it with unfailing regularity, urged on by the knowledge that it was expected of him, and resolutely undeterred by the thought, which must frequently have crossed his mind, that I hadn't the faintest idea what it all meant. I remember thinking that 'martial kind' must be some senior army officer of benevolent disposition, like Marshal Foch, a name familiar even to a small child so soon after the First World War. As for the meaning of 'enterprise', I'm pretty sure I hadn't a clue.

'Gilbert and Sullivan, selections from' also dominated our small repertory of gramophone records. I can still see the red label with the little dog on it, revolving at seventy-eight revolutions per minute, with a hissing noise and a haunting smell on the polished wooden machine which was kept on the second shelf under the dresser in the dining-room as the Band of the Coldstream Guards churned out a medley of airs from *The Yeomen of the Guard*. Certainly we had a few recordings of works by other composers. *Carnival of Venice* introduced me to the idea of 'theme and variations', while *Sweet and Low* and a couple of negro spirituals acquired a special significance because they came in a little album with brightly coloured illustrations of their respective lyrics.

An important milestone was passed during one harvest festival. In the late summer or early autumn of every year, when the church was filled with the pungent odour of fruit and vegetables, the Stibbard Brass Band (later to be elevated to the status of The Stibbard Silver Band) was invited to supplant Miss Boulter, whose job it was, as well as keeping one of the two village shops, to provide the music throughout the rest of the year on the harmonium. The licensee of the King's Arms collected his little group of musicians at the back of the nave and I was allowed to go and stand beside them. I don't know how old I was at the time, but old enough to know the tunes of the harvest hymns. The revelation, for it was no less, came in the hymn *Come ye thankful people, come*. In the fifth line, where the melody pursues a little rising sequence (Figures 42a and b), the tuba suddenly ran amok and headed in the opposite direction like some maverick demon tip-toeing down the stairs to the nether region. The effect was stunning! I could hardly wait to tell my mother what I had discovered, namely that the man with the biggest trumpet of all had started playing a different tune from everybody else. Such was my introduction to the twin concepts of harmony and counterpoint.

Twice in my life my mother decided that it would be a good thing if I learnt to play an instrument myself. Before I went away to school I began to learn the piano with Mrs Hayhow, a farmer's wife in the village. She was a good teacher and I liked her, but lessons soon had to come to an end when I went to the Glebe House, and I readily agreed with my mother that I should probably have enough

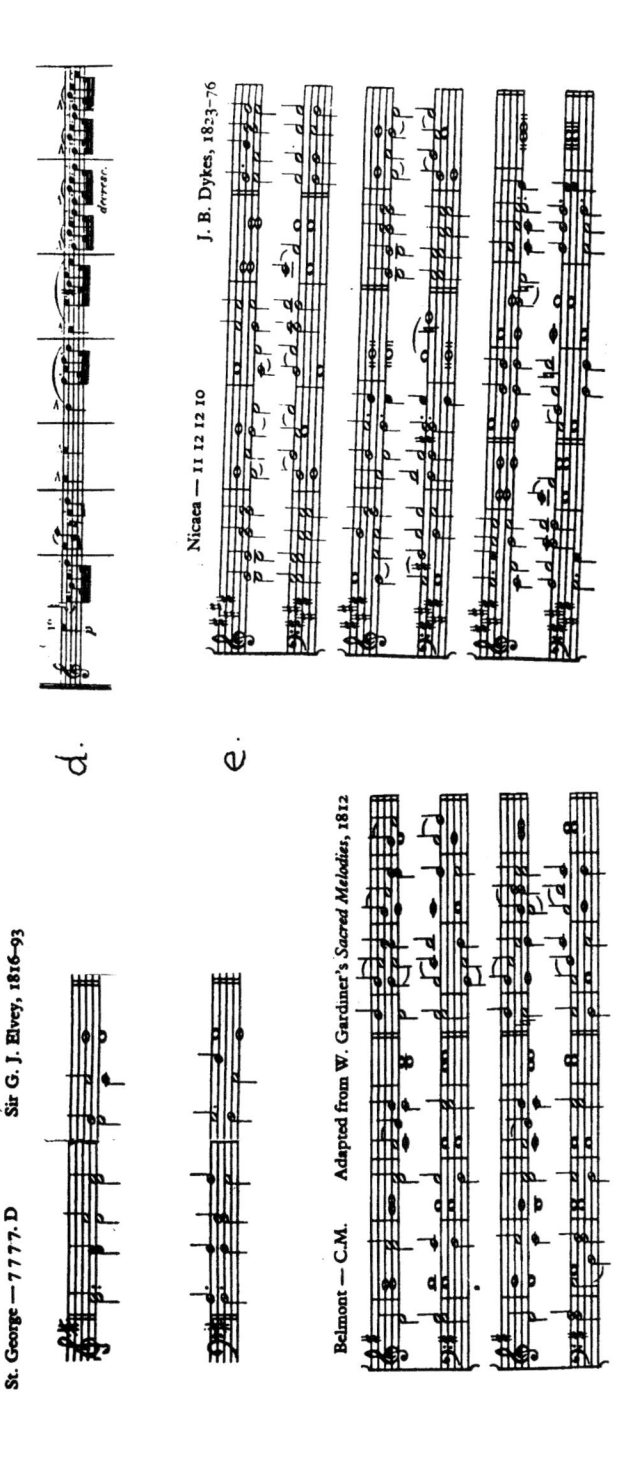

FIGURE 42. MUSICAL MEDLEY. (a) and (b) The fifth line of Elvey's tune to the hymn 'Come ye faithful people come' (*Hymns Ancient & Modern*, No. 482); (c) Belmont (*Hymns A. & M.*, Nos. 189 and 515); (d) Oboe theme from the Second Movement of the *Great C Major Symphony* (Schubert), and (e) John Bacchus Dykes' tune to the hymn 'Holy, holy, holy' (*Hymns A. & M.*, No. 160).

to contend with and that it would be reasonable to defer my lessons for a year. The year became two, then three, but eventually, in my last year, I resumed under the tutelage of Dr Bone, the King's organist from Sandringham, only a few miles down the road, who used to come over one day a week to give lessons at the school. I liked him too, not least because he was the only grown-up I ever heard speak injudiciously about the Headmaster's shortcomings; but progress was slow, and, when at the end of the year my mother again found herself concerned that I should have plenty to do in my new surroundings at Shrewsbury, I was once again happy to share her concern.

That, then, was the end of my instrumental tuition, apart from a brief if valiant attempt by Iris; but she was a good musician, and I dare say I was discouraged by the width of the gap which separated us, and wartime conditions were not the most conducive to regular practising. I think in retrospect I might have fared better had I attempted another instrument. I don't seem to have the sort of brain that can cope with several notes at the same time. The human voice, however, is a single-note instrument, and, though I've never had a good one, I was persuaded, on arriving at Shrewsbury, to join the chapel choir where I eventually became principal treble.

My persuader was the newly-appointed director of music, J. Barham Johnson, who immediately secured my co-operation by disclosing that he was a fellow East Anglian. We invariably conversed in the Norfolk dialect, and the interest which I had begun to acquire in serious music before I left owed more to him than to anybody else. When my voice broke I joined the basses in the school choir and it was there that I had the opportunity of learning some unusual works thanks to Johnson's predilection for those composers who had been active during his student days but by this time were less fashionable. *Hiawatha's Wedding Feast* is still heard occasionally, but how many singers more eminent than I have committed to memory the bass choral part of Stanford's *Phaudrig Crohoore*? In the House Singing Competition I learnt to sing madrigals, and in later years Iris and I have derived much enjoyment from singing in the Hull University Choir.

The seeds of my musical taste were sown, I suppose, in infancy, and I've never lost the taste for Gilbert and Sullivan; but a vigorous re-seeding took place during my later years at Shrewsbury, partly through my involvement in singing but partly also through the gramophone which was the instrument by which I was introduced more seriously to the German Classics. The Music Society (or was it the Music Club?) was nominally run by the boys but with a good deal of help from Barham Johnson. By the time I left the school, or very shortly afterwards, I could recognise all the Beethoven symphonies and at Oxford I spent a good deal of time extending my knowledge of Mozart, Haydn and Bach, in short any composer with a German name who wrote before 1830. I explored with pleasure the works of the Italians and of later composers of various nationalities, but by my early twenties there had emerged a kind of hard core of Germans and Austrians who had become for music what Scotland was for landscape, the ultimate criterion of excellence. At that time it looked as though my taste in music might be destined to be almost as unadventurous as my taste in food, but in subsequent years the boundaries of my interests have been greatly extended leaving some large pockets of prejudice within a pretty catholic field.

I think it would be true to say that any piece of music which doesn't have a recognisable melodic line or which strays for more than a few bars from the harmonic system of the major and minor key fails to interest me. Listening to such a piece is like reading a letter in a language I can't speak. Both intellectually and emotionally it's a non-event. It makes sense to me to see music as a kind of alternative communication system analogous to the spoken or written word. I don't mean that a piece of music must have a meaning in the sense of a programme appealing rationally to the intellect, but it must have some kind of shape, of individuality; perhaps *imageability* is the right word. Irregularities may be admitted, but only to give a greater significance to the undercurrent of regularity which must always be there as a kind of baseline of reference. Discords may be introduced, but only for the pleasure which comes from their resolution; syncopation may be fascinating, but only because it injects irregular beats within a continuing regularity of rhythm, while changes of key simply emphasise the wider framework of a harmonic system within which everything worth listening to is incorporated.

Compositions which fail to conform to, or, worse still, which appear to mount a deliberate attack on this concept of an underlying musical order, are rarely able to command my attention for long. On the other hand, I can derive great pleasure from a very wide range of styles as long as they don't transgress these basic rules. Thus compositions of the twentieth century are in general less likely to interest me than those of the nineteenth, but Elgar, Carl Orff, Scott Joplin and the Twelve-bars Blues of Lower Basin Street are capable of taking their place alongside Brahms and Dvořák, Sullivan and John Bacchus Dykes.

It may well have crossed your mind that we appear to have made another deviation which takes us well away from the subject of environmental perception and landscape taste, but the burden of my message is that all human experience is inter-related, and, if we now rejoin the main path of our journey, I think you will see why it has been necessary to make this detour. There is, I believe, a valid comparison to be drawn between the capacities of sound-patterns to produce distinctive individual images and of visual patterns to suggest personalities in the way I have earlier described. It's this audio-visual link which, for me, introduces the landscape theme, and which goes on to raise the question 'If I build, not one world of my own, but two, one in sounds and one in pictures, what is the relationship between them?' I think it would be no exaggeration to say that most of the music I enjoy has the capacity for calling up landscape images in the mind's eye. Equally I'm sure that, of all the parameters of music, tone, texture, orchestration, pitch, etc., it is the recognisable individuality of the melodic line which is crucial in determining whether this kind of association can or cannot be made, but I am, as you will understand, speaking of my own personal experience; yours may well be quite different.

Very often the reason which lies behind such an association is quite elusive, but there are certain circumstances in which we can recognise plausible explanations. We must all be familiar with that kind of association which links a particular melody, air, subject or whatever we call it, with a specific place. From a vast number of such cases I'll choose a few examples to illustrate the two categories into which they seem to fall. In the first are to be found those associations which are general, that is to say are instigated by some commonly understood cause, and it's these which are most likely to be shared by other people. There must be many, for example, for whom the music of Elgar conjures up pictures of the Malvern Hills because of the known connection between the composer and this particular place. Folk or dance music which is characteristic of the cultural traditions of a particular country naturally brings to mind its landscapes. I imagine most people hearing a yodelling song would be reminded of the Alps. An enormous number of compositions also bear the name of some geographical location, and as soon as we know the title of the composition we make the appropriate visual link, provided, of course, that we understand the allusion. Sibelius' *Karelia Suite* will only conjure up a vision of endless forests and lakes if we know where Karelia is and what it looks like. There's no mystery about this and examples are scarcely necessary.

In the second category are those associations which arise from purely personal incidents. Let me give you two or three of my own. It is as certain that you'll be able to match them with yours as it is that they'll be different. I remember as a teenager, when my interest in music was beginning to dawn, acquiring a record of an instrumental arrangement of selections from *Carmen*. I knew nothing of the story or the words except that the opera had something to do with bull-fighting. I played it over and over again until I decided it was time for a breath of fresh air, so I cycled the short distance from Stibbard to a wood in Fulmodeston which backed on to the wood at Swanton Novers where we used to pick the lilies-of-the-valley. As I walked down the rides and through the clearings these tunes were still ringing in my ears, and there was forged a visual link which has survived for half a century. *Auprès des ramparts de Seville,* notwithstanding its explicit title, conjures up for me a picture, not of an urban fortification in Southern Spain, but of a coniferous plantation in North Norfolk.

A powerful association of this kind concerns a hymn-tune called *Belmont* (Figure 42c) which for over thirty years has possessed the capability of evoking a picture of a small, barren island with a gleaming white lighthouse in an expanse of blue sea. I can date the association precisely to 11 o'clock on the morning of Sunday, 26 February 1961, when it was played over the ship's loud-speaker system as the P. & O. liner *Himalaya* rounded the little Perim Island in the Straits of Bab-el-Mandeb at the entrance to the Red Sea. For a shorter period, since 7 November 1972, the Schubert Octet, which I already knew well, has been mixed up with the little hills around Pannonhalma in Hungary. It was being played on the radio of a Government car in which I visited the monastery there under a cultural exchange scheme. Personal incidents of this kind can over-ride what might appear to be more probably associations. For instance, one might expect Russian Orthodox church music to suggest images of the Russian landscape (as does the theme tune from *Dr Zhivago*, which I first heard when watching the film); but for me it's essentially Parisian, because it takes me straight back to the Russian Orthodox Cathedral in Paris where I first heard that kind of music on my visit there in 1939.

In all these examples the landscape which became 'twinned' with the tune was determined by the accident of where I happened to be when I heard it; but sometimes there is no such actual geographical link, the connection is made by the imagination alone. Among the records of popular operatic airs available in the middle and late 'thirties I had a number of selections played in orchestral arrangements by Marek Weber and his Orchestra. This was my introduction to most of the operas of Verdi. As with *Carmen* I knew little or nothing about the plots or locations of the stories. Had I been better informed I suppose *Aïda* should have set in train a kind of Egyptian reverie with pyramids, palm trees, scenes of the desert fringe and the life-giving Nile. All I knew, however, was that Verdi was an Italian composer, so the landscape images which the music evoked contained a vast flat foreground with lombardy poplars, which is how I imagined the North Italian Plain, and a background of lakes and wooded hills with, in the far distance, a skyline of snow-capped Alpine peaks. The scenery of *Aïda*, in short, was the same as the scenery of *Rigoletto*, *La Traviata* or *Il Trovatore*, intensely evocative and extremely beautiful, but wholly unconnected with the stories of the operas themselves. Later familiarity has partly, but only partly, replaced these associations by others more relevant to their respective plots.

There's one other kind of link between melody and landscape which I'll mention because it ties up with something I've already told you about and illustrates the extremely individual character of these associations. Certain musical themes have for me their own consistent associations with the points of the compass. I've just mentioned the standard Verdi panorama with the Po Valley in the foreground and the Alps beyond. This implies a view approximately to the north and follows the first of the three principles explained above (p. 120). More commonly these 'musical views' follow the principle exemplified in Washington D.C., which, as you may remember, is always perceived as if it were being approached from the direction of my home. By the same token all the landscape images which fill my mind when listening to the music of Sibelius, including the *Karelia Suite*, appear to be perceived from the south-west, because Finland lies to the north-east.

Sometimes these directional associations, though no less strong, are more difficult to account for. For example, the little oboe theme in the second movement of Schubert's *Great C Major Symphony* (Figure 42d) also has a powerful 'north-eastern' feeling, though the picture here is not so much a landscape, but rather a skyscape of distant cloud images. *Per contra* John Bacchus Dykes' tune *Nicaea*, which is generally sung to the hymn, 'Holy, holy, holy', (Figure 42e) has always been equally evocative of a view to the south-west. Quite recently a possible explanation occurred to me when I thought about the line 'Casting down their golden crowns around the glassy sea'. The afternoon sun reflected on the water would imply a south-westerly direction (in the northern hemisphere). But it's safer and more honest to admit that, if I were asked to explain the origins of this kind of phenomenon, in the majority of cases I should be quite at a loss.

I'm pretty sure that most people, reading my account of these directional associations, would find the whole thing a load of nonsense. At the same time they would, I believe, respond to the suggestion that auditory and visual images are frequently linked, even if in ways quite different from those which I've described and even if none of us can understand the mechanism which links them. Simple as the idea is, it's extremely difficult to put into words. I did once have a shot at it, and here, for what it's worth, is the result:

SUMMER MOOD

Once in the cloudbed of a summer day
I saw, or rather sensed, a little mood;
And there he played above the hollow wood,
Borrowing feathers from the fluffy sky
To make an ambience, and by-and-by
Whispered a mystery, and slipped away.

How many skies have stretched themselves again
To grow warm colours in the creamy light?
How many clouds as downy and as bright
Have trailed frail plumage through hot summer days,
Slowly dissolving in the heavy haze,
To tempt my little mood? But all in vain!

I learnt this afternoon how Dvorák sought
In foreign clouds that same elusive mood,
Chased him above some deep Bohemian wood;
And how the fugitive that sultry noon
Assumed the likeness of a haunting tune,
But failed to slip his captor, and was caught.

So when I search a summer sky and see
No traces of my faithless mood at play,
I'll whistle him that dreamy little lay,
Plaintively mirrored in a minor key,
Which lovers use to mourn inconstancy,
And then - who knows? - he may come back to me.

FIGURE 43. LOOKING UP IN NEW ZEALAND. The Fox Glacier seen from its terminal moraine. Photo by the author, 1980.

FIGURE 44. LOOKING DOWN IN AUSTRALIA.
Richard Appleton as a teenager contemplates the gorge below Ebor Falls, New South Wales. Photo by the author, 1961.

7 MY KIND OF LANDSCAPE

'I had a world about me - 'twas my own;
I made it, for it only lived to me.'

I've tried to lay before you a selection of the anecdotes, episodes and experiences which seem to have something to say about the development of my habits of environmental perception, though the relevance of many of them may not yet be apparent. One dimension of the phenomenon of environmental perception concerns our aesthetic reactions to what we perceive, our likes and dislikes, our preferences and our tastes. In general, as might be expected, my tastes would be shared by a large number of people, even those with very different backgrounds, but within that general framework I've acquired a number of particular kinds of preference, some of which deviate quite markedly from the usual trend. I've little doubt that you've done the same.

Sublime Places

I suppose most people would find themselves powerfully attracted towards the stupendous. In the eighteenth century it was called The Sublime and it encompassed anything which was large enough, strong enough or impressive enough to induce a feeling of awe. Mountains have captivated me from those early days when I first encountered them in the Lake District. Modest as were the Cumbrian hills compared with the high peaks of many other countries they were so much bigger than anything we had in East Anglia. The far larger mountains which I've seen since have displaced them as the criteria of sublimity but not, I think, of charm.

I've heard it said that for the finest scenery in New Zealand one looks up (Figure 43) while in Australia (Figure 44) one looks down; the depths can be as sublime as the heights. I became acquainted with the great chasms of Eastern Australia when I lived for a year in

Armidale, New South Wales. This little city stands at an elevation of over three thousand feet on the New England Plateau, part of that huge tract of high ground which forms the misleadingly-named 'Great Dividing Range'. There are places, as in the Snowy Mountains, where the 'range' assumes something like a mountainous form, but there are other places, some of them near Armidale, where the eye can't detect the location of the actual watershed, so flat is the surface. On this plateau the rivers rise and gently meander eastwards in the shallowest of valleys to be suddenly confronted with precipitous drops of a thousand feet or more into forested ravines which hustle them away unseen in funnel-shaped valleys from which they eventually emerge into the coastal plain.

Several years were to elapse before I was to see the Grand Canyon of the Colorado River. If the saying is true of anywhere that no photograph can do justice to the original, it must be true of the Grand Canyon, which consists of a ravine of huge proportions within a whole system of vertical walled valleys in all four thousand feet deep and over ten miles in width from rim to rim. It's the concept of horizontal distance coupled with the depth of the chasm which creates the amazing feeling of immensity. The gorges of Eastern Australia are quite small by comparison, but, because they tend to be so narrow in relation to their depth, they're scarcely less awe-inspiring. While I was living in Armidale the capacity of the imagination to sense a landscape beyond the visual limits imposed by the horizon kept alive a perpetual awareness of these plunging cataracts somewhere to the east, between me and the Pacific Ocean. It was very like the feeling engendered by living on The Site at Shrewsbury (p. 68) only more so.

Not just the drop but the plunging water also is a common source of satisfaction. I became fascinated by waterfalls at an early age, and a waterfall would always provide a good picnic site on a long journey if there wasn't a railway available. During childhood I had to be content with the waterfalls of Britain, which were modest but nevertheless exciting. When much later I saw Niagara and Victoria they were indeed sublime, though Niagara, for me, inevitably suffers to some extent from the fact that so much of the immediate environment of the falls is now man-made (Figure 45).

At the Victoria Falls (Figure 46) it's true that the evidence of the hand of man isn't far away, but it's almost entirely out of sight. Even the railway bridge which spans the gorge below the falls is tucked away round a right angle bend in the Zambesi River and is invisible from most vantage-points. The Victoria Falls, therefore, afford the opportunity, which the Niagara Falls do not, to approach the spectacle through the natural vegetation, and that includes a small patch of rain-forest. If ever one needed to be reminded how the

FIGURE 45. NIAGARA FALLS FROM THE CANADIAN SIDE. Photo by the author, 1972.

FIGURE 46. 'THE SMOKE THAT THUNDERS'. The Victoria Falls from the Zimbabwe side. Photo by the author, 1975.

phenomenon of environmental perception is a product of the co-operation of all the senses, this is the place to seek a demonstration. One can hear the waterfall, feel it through the soles of one's feet, smell it, even taste it before one emerges from the forest to see the cause of the commotion. The sense of sight is the last to be rewarded.

Exposed Places

The term 'sublime', in the eighteenth-century sense, encompassed not only objects of great height and depth, not only manifestations of great power, but also anything which suggested extremes, for example of light and darkness. Not least a sense of the stupendous could be invoked by a sense of great distance. This has been so important to me that I'm not infrequently teased about it. I'm not interested, Iris tells me, in seeing anything unless it's too far away to be seen. I remember standing on the summit of Ben Nevis, the highest point in Scotland, on a very clear day and looking to the west where the line of the Outer Hebrides was clearly visible. Because it consists of a number of islands of clearly recognisable shape, differing in length and separated by visible gaps of open sea, it's virtually impossible to confuse their identity. The most southerly extremity is the cliff of Barra Head, and there it stood, crystal clear, just ninety-nine miles away.

Since then I have had other views as long and even longer, from Cobden on the West Coast of the South Island in New Zealand to Mount Cook, exactly a hundred miles away, from Point Lookout in New South wales to Mount Lindsay near the Queensland border, a hundred and fifty. From an aircraft even longer views are possible. I remember seeing the snow-covered cone of Mount Shasta in Northern California over two hundred miles away, and when I've been able to see identifiable objects on both sides of the plane simultaneously I've found it immensely exciting to feel the magnitude of the distance between them. On a flight from London to Budapest I could see the Hohe Tauern on the right and the High Tatra on the left simultaneously, that is to say from the borders of Italy to those of Poland, some three hundred and fifty miles apart, roughly as far as from Aberdeen to Bergen or Boston to Baltimore. Some of the satisfaction of such an experience comes from being able to identify what one is looking at, which is also a pre-requisite for calculating distances, and I'm probably better qualified in this than many people who are nevertheless 'turned on' by the mere awareness that what one is looking at, whatever it may be, is a hell of a long way away.

Satellite pictures have made possible views over much greater distances still, but the photograph, however, stupendous, can never quite match the reality of direct observation. For me much shorter distances can be highly evocative when the eye works in collaboration with the imagination, and never more so than in flat country where the level horizon invites the assumption that the land must go on and on and on. This was the feeling I first consciously experienced on the high ground above Madingley (p. 90) when a sense of exhilaration eventually triumphed over one of monotony and boredom which in childhood I had come to associate with the Fens.

A similar feeling has invariably been aroused when I've approached the Low Countries from England by sea. The knowledge that the whole continent of Europe lies behind those low fringing dunes invests the lowlands with a property which they may or may not possess. Images of rolling heathland and forest and eventually bold hills and mountains, seem to draw the imagination over the dunes, over the marshes, over the polders towards an unseen and certainly unseeable objective, far, far away. Some people have been attracted by the simplistic argument that, because mountains are widely acclaimed as beautiful, therefore flat country must be intrinsically devoid of beauty. Nobody familiar with the rules of logic would fall into this error, but even those who are not have to face the paradox that one of the greatest schools of landscape painting in the history of Western Europe, indeed of the World, flourished in the seventeenth century in Holland and Flanders, and any theory of landscape aesthetics which dismisses the attractive powers of flat country would have to take this into account.

One of the characteristics of flat country is that perpendiculars of even quite modest dimensions erected from the horizontal acquire a greatly exaggerated impression of height. In the Netherlands, for example, church towers and windmills assert a kind of authority which derives simply from the absence of any rival yardsticks by which to measure height. One sight which made a deep impression on me was that of occasional rows or clusters of lombardy poplars rising from the Great Alföld, the Great Hungarian Plain. Even where the foliage blotted out a part of the sky the spaces between the lower trunks allowed the eye to penetrate for twenty or thirty miles to what seemed to be an infinite distance.

There is, perhaps, a certain melancholy which easily attaches itself to flat places. I suppose that, in infancy, I found this alienating and depressing. Perhaps one requires the maturity of experience before one can find in it an emotional reward; but, once one has come to terms with the idea, the melancholy itself can be evocative in a very pleasant way.

Flat places are formed in one of two ways. Either they are the product of processes of erosion which have worn away higher ground until what is left is all of the same height, or they have been built up by solid matter being deposited, usually in water. A special kind of horizontal surface results, of course, from the freezing of water in lakes, rivers, or even the sea. The Dutch painter, Hendrik Avercamp, made his reputation out of painting them. Relatively flat landscapes can be produced in other ways, for instance by particles being blown by the wind or by certain kinds of flowing lava consolidating in near-horizontal surfaces, but by far the most common cause of absolute flatness, if it's not erosional in origin, is the in-filling of bodies of water, such as lakes or the fringing shallows of the coast, and an important visual significance attaches to the lines of contact where they meet the more elevated terrain which delimits them.

Sometimes such processes of infilling are incomplete. The resultant surfaces have a particular fascination for the eye, which discerns a single horizontal plane, part of which consists of land, with all the variety of colour and texture which clothes it, and part of water, which, while still engaged in the process of replacing itself with solid matter, continues to display its own contrasting characteristics, reflecting the sky and communicating to the observer an entirely different feeling, an entirely different mood.

The Great Plains of North America are rarely absolutely flat, but their vastness is such that small undulations can't destroy the sense of distance which is the overwhelming feature of the prairies. In Southern Canada, for example, west of Winnipeg, the prairie is trenched by shallow valleys which carry the larger rivers, like the North and South Saskatchewan, a hundred feet or so beneath what is really a low plateau rather than a plain. After completing my first crossing of this huge expanse by train at a time when the fall colours were at their best, and having reached the Rockies in Jasper National Park, I sat down on a rock and, repeatedly scolded by Columbian ground squirrels, tried to find words to express the sensations I had just experienced in the Montreal-Vancouver *Transcontinental*.

SEPTEMBER DISTANCE

This silvered eel, nosing from east to west,
Evokes a strange humility.
I never guessed
The world had so much space to spend on me.

Manitoba

Eastward, where birch and shivering aspen shone,
Beauty itself could never die,

Now these have gone,
But distance still survives to testify.

Nature, enslaved by man, and made to fill
Triumphant towers of blinding white,
Works with a will
To please her lord, who cringes out of sight.

But groups of exiles, whispering in the breeze,
Conspire to claim their lands again,
And shimmering trees
Creep back in clusters on the endless plain.
Saskatchewan

Although the streams below the prairie's face
Can never see the far-away,
They swallow space
Mile after sweeping mile to Hudson Bay.

Then seasoned woods with autumn-failing leaf,
Like Roman serfs about to die,
In golden grief
Salute the Sun-god in the ice-blue sky.

Alberta

Now the full forest claims its rightful place.
Alternate spires of gold and green
Strive to efface
The wild horizons of the distant scene.

At last the crumpled crests of blue and white,
Where the world ends in rock and snow,
Stretch out the sight
To lengths undreamt of in the plains below.

But distance cannot die, and even they
Taunt the imaginative mind
And point the way
To vaster distances which lie behind.

Enclosed Places

It may at first sight seem incongruous that anyone who can be so captivated by the phenomenon of distance and feel so strongly about open space should also find delight in enclosed, secret places. Some people, it's true, show a marked preference for one or the other type, but for me they are equally capable of inducing pleasure. How far, if at all, the delight in enclosed places was born under the rhododendron bushes of Bradley Hall can be no more than a matter of speculation; but I haven't the slightest doubt that familiarity with that private world of intertwining stems and branches, secluded from every prying eye, established the realisation that, however much one may be exhilarated by the wide open spaces, the shadows have a fascination of their own.

As I grew up I became familiar with other kinds of environment which possessed the power of evoking similar sensations. The most extreme of these was the rain-forest which I've encountered principally in Australia and Southern Africa. Where there's a superabundance of water and heat the controlling factor is light. Intense competition for light results in the plants reaching as high as possible to outdo their rivals and forming a dense canopy of leaves beneath which light starvation ensures a lack of foliage. The interior of the true rain forest is therefore something like a cathedral with tree-trunks for pillars. On first acquaintance it immediately recalled the darkness under the rhododendrons of Bradley Hall from which it differed chiefly in the much greater elevation of its canopy and the consequent feeling of spaciousness, but they were both dark and gloomy.

Darkness and gloom are not universally regarded as pleasant features of an environment, and in England today there is a widespread animosity towards the forests of densely planted conifers and particularly those of the sitka spruce. Attempts to rationalise this animosity often lead to their being condemned as 'exotics'. Their detractors brand them as 'not native', which is true, 'not natural', which is of doubtful meaning, or simply 'not English', when what they mean is that they don't like them. They complain, with reason, that they harbour a much poorer fauna than native English hardwoods. They object to the fact that they are often planted in 'unnatural' rows (like most of those trees which were planted in the seventeenth century for purely aesthetic purposes!) but I suspect that what they really dislike is usually the darkness and the gloom.

This is one of many areas in which I seem to differ from the majority. I too like a rich fauna; I prefer the more 'natural' appearance of irregularly planted trees (though I have already testified to a certain feeling for the straight rows in the orchards of Bramfield), and I find delight in the shafts of sunlight penetrating the more variegated colours of the hardwood forests especially in spring and autumn; but I find it difficult to get upset about the sitka spruce and the gloomy solitudes which lie beneath it. I find a similar divergence from popular taste in my attitude to the architectural counterpart of the forest to which I've just referred - the cathedral. Most connoisseurs of English ecclesiastical architecture speak with admiration of the lightness and airiness of our great churches, not least in those of my own East Anglia; but the churches which I find most memorable are often darker, more cavernous, feebly warmed by shafts of coloured light from stained glass windows or, better still, the glimmer of flickering candles. I wonder whether the dark interior of the oak room at Bradley may have had something to do with it?

If I have correctly perceived my own sensitivity to the gloom of the rain-forest, the sitka plantation and the darkened church, perhaps it should follow that I would enjoy even more the profound obscurity of the cave. Some of my friends are committed cavers and find in their hobby one of the most stimulating experiences of their lives. I can see the fascination and I can feel the excitement, but before I can become deeply involved in the sense of total envelopment which is so much more intense than that of even the darkest forest, other sensations come into play. What if one were unable to retrace one's steps? What if the sense of confinement were to become so overwhelming as to induce the feeling that one had no room to move freely, even to breathe? I long ago learnt that, just as my feeble urges to explore the sensations of eating unfamiliar foods were so easily extinguished by the fear that I might find them unwelcome, so my much more powerful urges to explore my environment reached a point where they were liable to suffer a metaphorical suffocation long before I was in danger of suffering a literal one. Claustrophobia isn't a serious component of my make-up, and I had ample opportunity to come to terms with it, such as it was, in the bomb disposal squad, but it comes into operation soon enough to deny me more than the most rudimentary pleasures of the speleologist!

Water and Sky

If there's one point on which almost everyone would agree it's probably that water has some universal aesthetic appeal. This protean substance appears in a wide variety of guises from ice and snow to the tiny droplets which form the clouds and render solid objects invisible in a blanket of fog. Even in the liquid form it's far from constant. Its appearance will vary greatly depending on whether it's still or swift-flowing. The gentle breeze which ruffles its surface transforms its apparent texture, destroying its capacity to reflect recognisable objects, changing its colour and inducing in the observer an entirely different mood. Stronger winds turn it from a symbol of tranquillity to an expression of those forces of nature which we think of as sublime when successive waves thump against a cliff, a sea-wall or, more dramatically, the side of a ship. While everyone responds in some degree to these aesthetic stimuli we each have our own individual quirks born of our own unique combination of experiences.

Some of my own images of water have been influenced by the impact of my fascination, to which I have earlier confessed, in fishing. Although I had virtually given this up by the age of twenty there are certain conditions, certain kinds of view of water which still evoke associations that I've no doubt were initiated during that period. For those who like to think of themselves as 'proper fishermen', that is to say those who cast a fly on a salmon river or a trout stream, I suppose the sight of deep pools alternating with white rapids must call to mind a whole way of life in which their most intense pleasures are to be found. I can understand this, and on one occasion in my late 'teens I visited a cousin of my father's, one Robert Pashley. Having emerged from the First World War as a partial invalid, he found himself with so much leisure on his hands, which he spent almost entirely on fly-fishing, that he regularly broke all records and earned for himself the title of 'The Wizard of the Wye'. Interestingly enough he was by then spending more time with his camera than with rod and line, challenged by the immediacy of the response required to train the camera on a leaping salmon and take the picture before his silver subject fell back into its own secret world.

But my kind of fishing had been very different. It was pursued in an environment of placid East Anglian waters, slow-moving streams, motionless lakes and even stagnant ponds. On their banks I would stand, sit or even lie on my tummy for hours on end trying to outwit creatures which seemed to employ every ounce of ingenuity in exploiting the strategic advantage afforded to them by a mystical aquatic environment, in some ways like my own, in others so very different. So the sight of slowly moving water, as seen from a bridge, sweeping the water-weed aside to open up deep glimpses of the river-bed beneath, invariably invites the image of a world fit for heroic fish to live in.

There is one particular variation of this general picture which seems to have a special capacity for conjuring up this kind of nostalgia, because I'm sure that's what it is. It consists of a flat meadow, grazed short, but still bright green under a winter sky, grey or blue, it doesn't matter. The surface of a stagnant pool almost level with the surrounding grass, is ruffled by a breeze stiff enough to render its surface totally opaque. Strangely enough the very denial of visual access to what lies beneath serves to stimulate such an intense curiosity that the sense of penetrating that fishy habitat is almost as real as when the eye can reach it without interruption. The details may vary but as long as the general picture is maintained so is the feeling. If, however, the banks were sloping steeply, this particular magic wouldn't work.

Mention of grey and blue skies brings me to the subject of clouds. Although I share with most people the feeling that the world looks at its best under a bright sun shining out of a clear blue sky, cloudy skies of various kinds may afford their compensations. Isolated cumulus clouds, rising like towers above a landscape, appear as discrete objects, taking their places alongside other objects, like mountains, trees or buildings, in the totality of the picture, as many landscape paintings will show; but even when clouds coalesce into a continuous blanket, obliterating the sun and losing their individuality as clouds, they don't necessarily frustrate the aesthetic effect, particularly if visibility isn't impaired by dust or moisture in the atmosphere.

I remember just such a day in the French countryside a few miles south of Paris in March 1939, on my first visit abroad. It was a Sunday morning and the air was exceptionally clear under a continuous canopy of light grey stratus cloud, and it's this picture which is brought to mind whenever I encounter such atmospheric conditions, no matter where the place or what the time of the year. Late autumnal days with dry winds desiccating the crinkly brown leaves are delightful when the colours are kindled by shafts of sunlight, but there is a certain autumnal authenticity about long distant views perceived under a ceiling of grey.

Very rarely does a continuous cover of stratus cloud present a completely homogeneous surface. At least there are usually patches of lighter or darker grey for the eye to focus on and hence communicate a sense of relative distance, and where the cloud cover becomes properly broken as in fracto-stratus cloud with conditions of clear visibility prevailing below the cloud base, some exciting compositions can result. The clouds themselves once again become individual objects demarcated by sharp and clearly-defined outlines, and differentiated from each other by distinctive shades of grey, ranging from almost-black to almost-white. Nearly always such clouds are being carried along in a strong wind and this helps to impart that hint of animism which, as we have seen, is easily suggested by movement.

Vegetation

I suppose for most people the first association suggested by the word 'landscape' is with a combination of relief and vegetation. For me too, though my few experiences of the true desert have taught me that even places devoid of vegetation can be exciting, even sublime.

'Vegetation' itself is a word which covers a vast range of phenomena from short fescue grasslands to immense forest trees. In all probability it was among the glacial and wind-blown deposits of North and West Norfolk that I first realised that many of the places I liked happened to be on sandy soils; Pretty Corner, Breckland and the very different, highly contrived but equally fascinating pine plantations in the dunes of Wells and Holkham. William Cobbett, I later learned, was contemptuous of the infertile heathlands, but in my scale of landscape values they have always rated very highly. It's not the sandy soils themselves that give me so much pleasure; it's the vegetation they tend to carry. A recent visit to West Norfolk after a gap of forty-five years or so brought bitter disappointment in places like Massingham Heath, remembered for their broad expanses of gorse and heather, with the occasional sentinel of silver birch, but ravished in the Second World War by the patriotic plough and converted into a fenced landscape.

Talking of fences, this brings me to one further confession of a deviation from the generally accepted aesthetic norm. Although I have been deeply fond of hedgerows ever since those early exploratory forays in Stibbard, and would generally regret their disappearance, I find it hard to side with those who seem aggressively determined on principle *not* to like landscapes from which hedges have been removed. Large expanses of open land do have a complementary role to discharge, and I remember being fascinated by the huge fields which the arable farmers had just begun to create in North-west Norfolk in the nineteen-thirties, harbingers of the age of large-scale agricultural machinery which was to come after the war. I shall get into terrible trouble for saying this!

A major reason for the opposition to the removal of hedgerows is clearly that it will impoverish wildlife by destroying cover and particularly nesting sites. This is a cogent argument, and as a conservationist I applaud it, but it's often urged as if it were the same as the argument for retaining hedgerows as landscape features on purely aesthetic grounds, whereas the two cases, though often complementary, are quite different. Many protagonists have weakened their case by failing to maintain this distinction and in consequence have succeeded in conveying the impression that they haven't really thought through the underlying principles. I believe a part of the antagonism to hedgerow removal arises from the fact that the resultant large open spaces are generally arable, and ploughed surfaces don't invite access in the same way as grass surfaces. Wide expanses of unfenced grassland rarely generate hostility, and indeed the intrusion of hedgerows to break up golf courses or parkland would almost certainly be condemned on aesthetic as much as on practical grounds.

Industrial Landscapes

Nor is this the only area of landscape aesthetics where awkward ambivalences have proved disturbing for me. Consider, for instance, those industrial landscapes of South Lancashire which gave me an early introduction to some of the worst areas of visual blight in England. Even at that age I could recognise them as the expressions of something radically wrong, what we would now call social and economic deprivation. In those days we called the disease 'poverty' and its symptoms 'the slums'. That there might be something in this landscape that one could actually like was an idea from which one felt one ought to recoil on moral grounds, as most people did, but to myself I couldn't deny that some of its manifestations had a certain compelling fascination. The Paleotechnic industrial landscape was far from being beyond redemption.

The grounds of Bradley Hall had been a little oasis of Edwardian gentility which had survived beyond its time, an oasis within one of England's great industrial landscapes. Its immediate environs were by no means unattractive in a conventional 'rural' sense. Beyond the rose garden a meadow stretched down to a tiny stream flanked by wet marshy ground, where we could pick wishy-washy bog-loving flowers for my grandmother, sheltered by the overhanging branches of an oakwood. Within this cosy haven nature still appeared to be in charge, challenged only by the occasional train riding high on the embankment which ruled off the horizon to the east; and who was going to complain about that?

But already there were more subtle, more clandestine intruders bearing the message that, beyond this little nest, the world had indeed been changed. The water in the stream had a murky quality and its margins were lined out by wispy strands of fine coal dust. The marks on shirts and jerseys brushed against tree trunks were not green with lichen or algae but dark grey with soot. The bluebells in the wood breathed a morning air well laced with the aroma of bitumen, and all along the field path that led back to Bradley Hall there protruded bits of old iron, broken pottery and torn fragments of half-rotted tattered textiles, tell-tale symbols which had infiltrated the oasis from the surrounding domains of the Wigan Coal and Iron Company.

Out there was the country where the real industrial landscape was to be found, towering chimneys, pit-head gear, cotton mills and, a little further away, blast furnaces. The whole landscape had been re-shaped by the hand of man. Before I was ten I could tell a slag-heap from a colliery spoil-heap; I knew about subsidence, even if I wasn't sure how to pronounce it, and I was rapidly acquiring a knowledge of where the most exciting sights were to be seen and how they were linked together by the complexities of a fascinating system of intersecting railway lines. For me the appeal of railway landscapes was not confined to beauty spots, like the Scottish glen which I've told you about. In Glen Garry the railway might be seen as an incongruous intruder, however welcome to me personally; but in the urban conglomerations of South Lancashire railways were the very arteries of an integrated industrial whole. There was a special kind of excitement to be found in their exploration, and by my late 'teens I was to spend much time searching them out as they crept through calandestine cuttings and coal-black tunnels in the inner approaches to town centres all over the country in the struggle to gain access to station platforms. The environs of stations like Bolton (Trinity Street), Liverpool (Lime Street) and Sheffield (Midland) for example, seemed to pose almost insuperable challenges, which the ingenuity of the railway engineers had nevertheless managed to overcome.

Later I was to explore all the industrial regions of Britain and to discover that the common characteristics of its collective image found expression in highly diverse and individualistic landscapes. By the age of ten I had assimilated that incomparable view from Granty's attic in West Hartlepool. I had seen half-constructed hulls being assembled in the shipyards of the Tyne and the Wear. Not least I had seen miles and miles of terrace houses destined to be immortalised decades later in *Coronation Street*. Not until I became aware of the social cost which had to be paid, not by me but by the people who had to live in it, did the magic of industrial Britain begin to wear thin and some quirk of conscience tell me that the world ought not to look like this.

Whether I would have found the addiction too strong to give up I can't say. The problem was solved by the acquisition of a more specific interest in the *relic* features of the industrial past. Long before the invention of the term 'industrial archaeology' I found myself fascinated by the remains of mines, furnaces and, needless to say, railways and canals, especially when they occurred in places where the landscape showed signs of reverting to wilderness. Just as fortresses, castles and other kinds of military architecture seem to acquire a romantic charm in dereliction which they never possessed in their active operational lifetimes, so the premises of mining, manufacture and transport acquired a new beauty in decay as nature wrapped herself round the fragments that survived, clothing them with ivy, filling their hollow spaces with birch and sycamore and re-roofing them with canopies of green foliage. Whatever my kind of landscape was, it had somehow to make room for all this.

The industrial landscape exemplifies a truth which is of general application in most inhabited landscapes, namely that most of what we see around us is not a manifestation of some order deliberately created to afford aesthetic pleasure; far more often it is the incidental product of some economic activity which has involved making visible changes in the natural environment. The hedged fields may please the eye, but that was never their purpose. They were made to facilitate the management and breeding of livestock. The forests of sitka spruce were not planted to change the visual aspect of Scottish moorlands, whether for better or worse, but to provide a source of commercial timber. The reservoirs, which have so often given rise to a love-hate relationship with the public, were designed to furnish a water-supply for cities or for the provision of hydro-electricity or both. Their designers were not primarily concerned with the paradox that the same people who had so vehemently opposed their construction could be seen, twenty years later, flocking in large numbers to worship them as beauty spots. Factories were built to make things, and even the most prestigious buildings, like city halls, were designed primarily to accommodate the business of civic administration. If they caused visual enjoyment, as they often did, that was a spin-off. But some landscapes, like some buildings, were designed primarily to give pleasure to the senses, and particularly to the eye.

Designed Landscapes

The contrived landscapes of this kind which I encountered in my youth varied greatly in size, in the degree of their formality or informality, in the balance between enclosed and open space and in many other respects. If I'd stopped to ask myself what sort of parks and gardens I liked I suppose I'd have had to be content with listing particular places for which I had a preference. The idea that there might be some underlying *rationale* which could form a basis for some sort of explanation of landscapes as aesthetic phenomena came to me much later. Even so I think I would have subscribed at quite an early age to the notion that designs which looked more natural were likely to be the more appealing, to me at least.

The influence of the urban park on my developing taste was less than it doubtless would have been had I been brought up in a town. I made the acquaintance of Mesnes Park in Wigan as the place where one waited in a state of some anxiety to go to the dentist. Most other urban parks were seen behind iron railings from a passing vehicle. Regents Park, as far as I was concerned, was an appendage of the London Zoo. I had few intimate experiences, either, of the parks which surround England's stately homes until I was a teenager, which perhaps explains why my encounter with Holkham was such a powerful emotional event and why, on my own doorstep, the grounds of Sennowe were able to make such an impact. It must be remembered that in those days the public was still rigidly excluded from many of the famous parks and gardens which today attract large numbers of visitors. So my exposures to these large-scale works of art, which used as a medium those very raw materials, trees, grass and water, to which I had already discovered I was romantically sensitive, came as something of an eye-opener.

Gardens of the smaller, more private kind played a central role in my early awareness of the world around me, but I was only marginally interested in them as places for the cultivation of flowers. Their virtue lay in the opportunities they afforded for exploring, which included such activities as tree climbing, and for assembling in the mind comprehensible pictures of their layouts; but I remember being aware quite early of the sense of seclusion which a garden can offer, a feeling I encountered, for example, in the gardens at Stibbard and Bradley and in my Housemaster's garden at Shrewsbury. The presentation of cultivated plants was less important than the creation of a mood.

Down Under

I think you will have realised that, with a few exceptions, most of the specific examples I've chosen to illustrate these facets of my landscape taste have been drawn from Britain. This is because the most important formative period lay in those years before I ever left its shores. As my experience of the world has widened whole categories of landscape types, not known in Britain, have been accommodated within my preference system, occasionally causing important revisions of that system though never radically upsetting it. Following that first visit to France, Germany and Belgium in 1939 I had a long wait before the opportunity to go abroad recurred. From 1951 onwards I made a number of further visits to the Continent using family holidays to enlarge my personal experience of some of those places on which I had been cheerfully lecturing without having ever been anywhere near them; but not until I was forty-one did I venture further afield than Austria and Italy. It was therefore with an immense sense of adventure that I sailed for Australia early in 1961.

Since the cost of air travel was roughly twice that of a sea passage, we set out from Tilbury *en famille* on the P. & O. liner *Himalaya*, destined for the University of New England, Armidale, New South Wales, where I was to hold a Visiting Lectureship for a year. Richard was fifteen, Charles just four and Mark a mere four months. The idea that a four-week cruise (actually twenty-nine days out and thirty-eight back the following year) was to provide an opportunity for relaxation was quickly shattered. Richard made friends of his own age and established his independence. Charles, however, came out in spots off the Mouth of the Tagus and was whisked away to cruise the Mediterranean and the Red Sea in splendid isolation in the quarters set aside in the stern of the ship for passengers with infectious diseases. You can imagine him lolling in his deck-chair, his little feet still well clear of the ground, a stiff Ribena on the table at his side, watching the *fellahin* at work on the banks of the Suez Canal from a position of aloof superiority on his private deck. When he was re-cycled into the ship's nursery the population of tiny tots was already being fast depleted as, one after another, they were carted off to the quarters he had just vacated. Embarrassment prevented us from enquiring how many had contracted measles by the time we reached Sydney. Mark imposed the usual demands of a four-month-old baby and I suppose Iris and I shouldn't have been surprised that most of the books we had taken to read on the voyage remained unopened.

Nevertheless it was a marvellous trip and it gave us a little insight into the world of Kipling and Somerset Maugham, just in time! Very soon afterwards the air fare was, in relative terms, to come tumbling down and a sea passage to Australia was to become a luxury for those more affluent than I, if indeed it could be obtained at all as one service after another was withdrawn.

As for experiencing the landscape, the great advantage of going by sea was that it afforded opportunities for at least fleeting glimpses of a number of wholly unfamiliar types. On the outward journey we went ashore at Gibraltar, Port Said, Suez, Aden, Colombo, Fremantle and Melbourne before making a spectacular dawn entry through The Heads into Sydney Harbour. On the way back additional stops were made at Piraeus, Naples and Marseilles. Some people today are so sated with travel that the whole business becomes a ghastly bore. To me, however, who had waited forty years to satisfy a passion, even these fleeting glimpses went a long way to quench a prodigious thirst for visual experience, and the memory of those shore excursions remains vivid.

The much more protracted contact which I was able to make with the Australian landscape during a year's residence created an opportunity to acquire some familiarity with an entirely new type of landscape and a whole range of unfamiliar species of flora and fauna. Perhaps the most important thing this visit did for me was to close the gap between those romantic images of far-away lands (like Australia), which I had begun to build up from picture-books in the nursery, and the sensation of training my own eyes stereoscopically on actual objects in real places and thus bringing them within the reach of my personal visual involvement.

I was later to see higher mountains (I had, in fact, already seen the Alps), more voluminous waterfalls, more extensive deserts, bigger and more volatile volcanoes and vaster forests, but by the time I returned from Australia I had encountered at least some examples of all the major types of landscape feature to be found on the face of the earth except for the great ice sheets of the Arctic and Antarctic. By dividing these major types into sub-types it's possible to produce a kind of hierarchical typology of landscapes according to their visual characteristics, and many geographers have attempted to do just this.

Aesthetic Values

To apply a similar technique for distinguishing between landscapes on the basis of their *aesthetic* values poses much greater problems for many reasons of which three are perhaps paramount. The first can be illustrated by looking at a particular type of landscape for which I have already confessed a strong liking, namely sandy heathlands with associations of birch, Scots pine and so on, such as might be found on the Surrey Commons and in many other localities. Each individual example, however, possesses some characteristics which are not possessed by others. Some may be flat, some steep. Some may be bordered by the sea, others may be inland. Some may be extensive enough to comprise virtually the whole view, others may abut against strongly contrasting types. The number of possible permutations is infinite, and when we're considering the feelings aroused by such landscapes we may find great difficulty in establishing with any certainty how far, in any one view, our pleasurable sensations derive from the characteristics of heathland as such and how far from its other components.

The second reason follows directly from this. Most people, when questioned about their landscape preferences, will call to mind, not a class or category of landscape, but a particular place. When I first went to the School of Geography in Oxford one of the first ideas I encountered which made real sense to me was the concept, developed by the French geographers, that places, like people, can be thought of as having a *personnalité* (shades of animism again!); but whereas every person constitutes unambiguously one discrete unit in any classificatory scheme, the word 'place' raises all sorts of problems of definition, such as the scale on which we conceive it and the boundaries by which we define it. A 'place' can be a whole country or a town or the corner of a field or a position at the dining-table or a patch of rough skin on the back on the neck, and the concept of a place having its own personality is valid at every scale.

The third reason brings us back very directly to the central theme of this book, because inevitably the personality which we find emerging in any place is deeply influenced by our own individual perception of that place. Places we have visited only on fine days, or where we have particularly enjoyed ourselves, or places inhabited by someone we're in love with, are likely to acquire a bonus to be added to any intrinsic landscape qualities they may possess.

It's in this context, then, that I find myself asking, not merely what is my kind of landscape, what category of visual environment within a classificatory system is most likely to be productive of pleasure for me, but also whether there are particular places which are exceptionally attractive. Going one stage further, I ask myself whether there is any one place which I can think of as being uniquely my own in the sense in which Gray's 'rude forefathers of the hamlet' were locked in a lifelong relationship with a particular patch of ground, and I'm forced to conclude that there is not. Since my roots were lifted from Stibbard at the age of twenty I've never again felt that they've been re-established quite as firmly in other soil. I've now lived on Humberside for over four decades, longer than all the rest of my life put together, but it isn't yet the unchallenged claimant to the title of my very own place.

The fact is that, when I come to think of all the other places which might qualify for such a description I realise that, even confining the list to places in England, the pictures I have of them differ widely from the reality. If I were to try and describe them to you, I should present you with such a subjective picture gallery that you would scarcely recognise them. For example, very different towns like Truro, Chippenham and Cardiff in my system would all belong to one family, tied together by the ethos of the Great Western Railway, the very concept of which might well be meaningless to you. For other people quite different criteria could fulfil similar roles - football teams, the territory of favourite breweries, vernacular architecture, holiday reminiscences, the residences of relations, momentary glimpses under unusual atmospheric conditions. We all attach significance to such different things that it would be remarkable indeed if the pictures which emerged were anything but highly diverse.

It had long ago occurred to me to wonder whether there were any general principles which could be invoked to explain taste and preference in landscape. Many people had already attempted to enquire into precisely this question, but those theories which I had encountered didn't convince me, and I dare say a major reason why I didn't address the problem seriously myself until I had turned fifty was this matter of diversity of experience as between individuals. How could one reconcile so many variations of opinion with the existence of general principles of universal application? How I was eventually led into this area of investigation will be the subject of Part III, where we shall also look briefly at some of the paths which others, often working in disciplines very remotely related to my own, had begun to explore from different starting points. What excited me was the discovery that many of these paths seemed to be leading towards a common destination.

PART III FROM PRACTICE TO THEORY

8   THE FACE IN THE MIRROR

'In years that bring the philosophic mind'

It's remarkable how, when we look back on our lives, we can see that what we have thought of as rational decisions, carefully planned, turn out to have owed a great deal to the hand of chance. Very rarely can we attribute important changes to a single event, a unique episode, an isolatable cause, unless, for example, we're unfortunate enough to suffer some incapacitating accident so drastic in its consequences as to over-ride every other influence on our subsequent life-style. More often we can recognise a combination of circumstances, often apparently unconnected, which lead us to embark on some change of course. There may be one particular event which acts as a trigger, but it can no more adequately account for what follows than the assassination of the Archduke Franz Ferdinand can *explain* the outbreak of the First World War.

When, therefore, I decided to apply some systematic rational thought to the sort of issues which I've outlined in the previous pages, this wasn't a decision which I can attribute to any one cause or particular date. It arose out of a number of circumstances which might well be thought unimportant and irrelevant to the pages which follow, were it not that the course of my reasoning could no more be divorced from its antecedents than could my taste in landscape be interpreted without reference to the events of my early life. In other words, just as events and circumstances have influenced (if 'determined' is too strong a word) my tastes and preferences, so also have they predisposed me to look for explanations along certain lines which may well be different from those pursued by other investigators who may equally be presumed to be, to some extent, the prisoners of their own life-histories.

Subjectivity, in other words, is a characteristic, not only of the way we perceive the world, but also of the way we interpret our own and other people's acts of perception. When we put forward hypotheses to explain the phenomena we have been investigating, we can never rid ourselves of it entirely. Even if our work is based on meticulously observed data, accurately recorded and independently checked, the questions we have asked in the first place are bound to be affected by the character and limitations of the instruments we employ, that is to say our individual brains. There's no reason to suppose that the same instrument which leads to so many different habits of perception and of landscape taste, should suddenly begin to operate as if it were not the product of its own past. It follows, therefore, that, if we are to attempt to establish connections between the case study we have been looking at (my own selective autobiography) and what we may call for want of a better term 'general theory', we can't yet abandon altogether the autobiographical approach, and, as a starting-point, we need to take some further note of what had been happening in my career as a professional geographer which must clearly have influenced the development of my investigating instrument, my brain.

The Nature of Geography

When at the age of nineteen I was introduced to the academic discipline of geography, there was already much active discussion about the nature of the subject, about its definition, indeed, and not much sign of general agreement. This didn't matter to me as long as the topics I was expected to study continued to be those which interested me, and, with few exceptions, this was so. Although my period of residence at Oxford was curtailed by the war, I spent long enough at the School of Geography to come to the conclusion that, however the pundits defined it, geography was a discipline which concerned itself with *places*, and if I could find nothing in this field to capture my imagination and fire my enthusiasm, since 'places' had always meant so much to me, then perhaps I'd better ask myself whether I had any business to be in a university at all.

I suppose that, of all the aspects of geography which I was required to learn about, that which interested me most was what places were like, and not least what they looked like. Fortunately I discovered that the interests of most of my teachers, while they may have led into branches of the subject in which I found less to excite me, generally started with a visual concept of places. This applied particularly to that branch of physical geography known as geomorphology, which studies the processes of formation of the physical features of the earth's surface, and it was this which, at that stage, chiefly appealed to me.

I left Oxford knowing a good deal more (I could scarcely have known less) about academic geography than when I arrived, but my experience of the subject was fragmentary and rendered even more so by the hasty alterations to the syllabus necessitated by the introduction of the 'War Degree'. There followed a period of nearly nine years during which the elementary understanding of geography which I had acquired at Oxford served to illuminate many of the experiences which came my way, in the army, on the fruit farm, and, more important, in my general observations of the world around me. When, however, I arrived in Newcastle at the age of twenty-eight, I was well aware that my knowledge of the subject was at least as fragmentary as when I had laid my books aside at Oxford. Not only that, but the fragments themselves were not tied together by any system. I had no geographical philosophy.

Most of the attempts I had made to read the more philosophical literature on geography, on its purpose, its meaning, its methodology, had failed to strike a note of relevance to my own way of looking at the world. Few of the authors I had encountered seemed to have the same visual approach to places which had led me into the subject in the first instance. It was as though there were two kinds of geography, the serious, philosophical, literary kind, which reminded me of my classical studies and was for real scholars, and the straightforward, observational kind, which was for me. Not that this latter approach had been neglected by my teachers at Oxford, as those weekly excursions into the surrounding countryside testified; but there seemed to be an indefinable gap between those highly pleasurable experiences of an exploratory kind and the exacting, scientific discipline which, one supposed, was what the university expected of its pupils.

When I arrived in Newcastle I was nearly nine years older, certainly more experienced, in some ways even a little more mature and I dare say more receptive to any force which might be available to introduce what had been missing. 'Force' was, I think, the right word, because, when it hit me, metaphorically I hasten to add, the impact was immediate and irresistible, and it did more to change my way of looking at the world than any single experience before or since. It took the form of a German scholar, some ten years older than I, who had left the Reich in the middle 'thirties to escape from the National Socialist régime. M. R. G. Conzen, 'Con' to everybody, including his wife, had studied at the University of Berlin and arrived in England as a product of the German educational system. If anyone knows about the rigours of academic discipline it's the

Germans, but those whose works I'd previously come across seemed to have been carried up like Elijah into a kind of élitist cloud which had no place for me.

I discovered almost immediately that Con seemed to have the same basic attitude to geography as I had, in that all his enquiries started with visual observation. For him the landscape was a repository of information which it yielded only to those who knew how to ask the right questions. This much I had in common with him, and, having discovered a common starting-point, I found I had the confidence, in myself as well as in him, to follow new paths into the unfamiliar. His methods were far more exacting than mine, his insight infinitely sharper, his accumulated knowledge incomparably greater and his scholarship of a different order, but I very soon knew that I had found the link I was looking for between the world of everyday environmental observation, the world of the boy scout, and the world of the philosopher.

I dare say the most important lesson I learnt from Con was to recognise the respectability of the elementary. I had hitherto been in danger of falling into the trap of supposing that anything academically worthwhile must be highly complex, whereas in truth it's the function of the academic to reduce the complex to a level of simplicity at which it can be comprehended. He taught me, for example, that everything in what he called the 'cultural landscape' (initially a German concept, *Kulturlandschaft*, covering the whole natural landscape as modified by human activity) can be interpreted in terms of three aspects, its morphology (its shapes, patterns and distributions; what it's like), its function (what it's for) and its evolution (how it came to be like that). A simpler formula could hardly be conceived, yet it has proved over the succeeding years to be the most useful exploratory tool that was ever placed in my hands. Of course, it's only a starting formula. The problems do indeed soon become more complex, but as long as the chain of question-and-answer can be related back to those three aspects, and as long as one remembers that all geographical questions must ultimately be concerned with *places*, one has the rudiments of a methodology which made a great deal of sense to me then and still does.

Teaching and Research

When I went to Hull my contract stipulated that I was to engage in teaching, research and such administrative duties as might be prescribed by the Head of Department. As far as teaching was concerned I found myself responsible for a course in historical geography, which, no doubt under Con's influence, had by now become my favourite subject, another on the regional geography of

Europe, in which I lectured with authority on places I'd never heard of, must less visited, and another on the interpretation of maps. Other courses were added in subsequent years, and there were in addition tutorials, seminars, etc. I had no choice in these matters, so I counted myself fortunate that my boss, the late Herbert King, foisted on me only subjects which I found congenial.

When it came to my second obligation, however, the pursuit of research, it was expected that I would have the initiative to put forward proposals of my own, and I wasn't long in deciding to invest a hobby with the dignity of a research project; so I chose railways. Although Con would not have claimed any particular expertise in this field, his main interest being in urban morphology (and incidentally I acquired a particular interest in that area in which they both overlapped), it was clear to me that the kind of approach he had been following in his own work could equally well be applied to the study of transport systems and that the trilogy of morphology, function and evolution were just as useful as guide-lines for research. I've already made some reference to this work in Chapter 6 and noted the danger of allowing the emotions to interfere with rational investigation. Being as honest with myself as I can, I believe in retrospect that I managed to avoid this particular pitfall reasonably well.

For twenty years I pursued my researches in this field. I succeeded in discovering nothing profound, nothing which, with hindsight, doesn't seem fairly obvious (which I suppose can be said of a good deal of research), but I did have the good fortune, working at a rather low level of sophistication, to light on a number of very simple ideas which nobody had apparently thought of before. At that time very few people had done any serious work in this field, so it wasn't difficult to find something original to say about it.

Late in the nineteen-fifties I found myself addressing a conference which was also addressed by the late Sir Dudley Stamp, probably the most prolific geographical writer of his generation, and afterwards he invited me to expand my lecture and write a book for a series he was editing. This I duly did, but the other contributors to the series were less diligent, and I well remember my disappointment when he told me the project had fallen through. Fortunately I was able to find another outlet and *The Geography of Communications in Great Britain* was published for the University of Hull by Oxford University Press in 1962.

Although in retrospect I think the title, which had been chosen to fit into Stamp's series, was not as appropriate as it might have been, it clearly implied a much wider field of reference than railways alone, and a major part of its message was that each kind of transport medium, being governed by its own technical limitations, displays a different type of adjustment to the land surface across which it runs. To improve my knowledge of the canal system I spent a few pleasant days camping with Richard, then a young teenager, near Watford Gap on the borders of Northamptonshire, Warwickshire and Leicestershire, and I wrote a little paper on the transport system of that area which was much in the news at the time because of the construction of the M1 Motorway (Appleton, 1960).

The New Geography

During the nineteen sixties certain changes had begun to take place in academic geography, and by the end of the decade these had brought me, at the age of fifty, to a point where I needed to review my own position within it. In retrospect I can now see that this was perhaps the most immediate cause of my dropping my research in transport and turning more seriously to the study of the landscape. If I now direct your attention to what these changes were, please remember that my interpretation is also the product of my own brain and no more likely to be a statement of absolute truth than anything else in this book!

What is sometimes referred to as 'the quantitative revolution' has gradually overtaken most disciplines. In the pure and applied sciences this happened long before I was born, and now it seems difficult to conceive that anyone could have made any important contribution to any branch of science in modern times without a mastery of quantitative techniques. We know, however, that Darwin was bemoaning his own inability to handle mathematics and statistics at even a very modest level at the very moment when he was busy turning the biological sciences upside down in the eighteen fifties. A hundred years later the social sciences had either been 'quantified' or were about to be so, and by the nineteen eighties many scholars in arts disciplines found themselves relying on computer techniques to process the raw material of their research.

In geography, which came into English university syllabuses as a serious discipline almost within my lifetime, advanced statistical and mathematical techniques had been practised by some geographers well before the sixties, but during that decade they began to be regarded as more or less a *sine qua non* for anyone seriously embarking on a career as a geographer. I was never opposed to these innovations as such; indeed it was very clear that with the computer the new breed of geographers could successfully tackle questions which could never have been tackled before. Entirely new horizons were opened up, and how could one oppose that?

My misgivings were quite specific and were related to three areas of concern. The first was purely personal, a criticism of myself rather than of the train of events. Although I took steps to remedy my limited ability, I have never reached the stage at which I can handle complicated statistical material without an expenditure of energy quite disproportionate to the satisfaction I derive from the results. I rarely get excited about truth as expressed in numbers. That's my loss, of course, but in any walk of life we're more reluctant to put our energies into those enterprises in which we're conscious of being inadequately equipped and therefore inadequately rewarded. That way we reduce the risk of disappointment.

The second area of concern was more serious. It seemed to me that some of my younger fellow-geographers became so interested in the new quantitative techniques for their own sake that they were in danger of exchanging the old for the new, rather than supplementing the one with the other. In its most acute form it meant that they lost sight of the questions which the new techniques would have enabled them to answer. Research students began to choose their research topics not so much for the questions to which they wished to find the answers as for the opportunities afforded by particular techniques. This was a widely expressed anxiety among many geographers (not only of my generation) and led to the rejoinder that those of us who expressed it had simply lost touch. This may have been true, but that's a matter which future historians will be in a better position to judge.

The third area of concern was again more relevant to me personally because of the nature of my interests. Quite simply these new techniques, when first introduced, seemed to lend themselves more effectively to the solution of problems which had little to do with the *visual* aspects of the world around me. Paper after paper, article after article, appeared in the geographical periodicals which, if they made any reference to specific places, and most of them did, left the reader without even the haziest idea of what these places looked like. The very driving force which had propelled me into academic geography in the first place had been removed from the field of play leaving it free for what sometimes appeared to be virtually a new game.

In the nineteen sixties, then, academic geography seemed to me to be striving to turn itself under the influence of Logical Positivism into a self-consciously 'scientific' discipline, a move which opened up new possibilities for some lines of enquiry but held out little encouragement for others. Within the mainstream, however, there were some minor counter-currents. There were emerging, for example, some initial signs of interest in what was to become known as 'humanistic geography' which sought to strengthen the 'arts' base of the subject, and I began to develop a fellow feeling (though not necessarily an invariable agreement) with the protagonists of this movement. It might be misleading and even invidious to name names, but the sort of geographers I have in mind included Hugh Prince, David Lowenthal, Douglas Pocock, Denis Cosgrove and Stephen Daniels, all of whom, being resident in England and therefore more accessible than their American colleagues, I came to know personally.

A second counter-current took the form of a more articulate awareness of what had always been realised by geographers, namely that, for many purposes, what mattered was not so much what the world is like as what people think it's like, a change of emphasis which led a number of geographers to become seriously interested in the study of environmental perception. A number of particular lines emerged, such as hazard perception, and one of these began to veer in the direction of the study of landscape preference which in turn led on to the analysis of landscape value and landscape quality. As for my own particular fields of research, historical geography was only just beginning to feel the wind of change, but by about 1970 the geography of transport was clearly heading in the new direction. Here again opportunities were opened up for making discoveries which would never have been possible under the old order, but the measuring of 'flow-lines' and 'desire-lines' or studies of 'origin-and-destination' and 'journey-to-work' were not objectives which fascinated me. The morphological, landscape-oriented studies of railways, roads and canals, which had commanded my attention for a couple of decades, looked to be in for a thin time, and it was perhaps with a sense of relief that I acknowledged to myself that, if I had ever had anything worthwhile to say on this subject, I'd probably already said it.

Change of Direction

One episode is perhaps worthy of mention because I can see, again with hindsight, that it formed a kind of bridge passage, to use a musical phrase, between the movement that had just ended and the one that was about to begin, though at the time I didn't know what this was going to be. In 1968 the United Kingdom Parliament passed the Countryside Act which among other things set up the Countryside Commission. One of its powers was to approve (or by implication reject) applications by local authorities for grants from central government to convert certain categories of unused or derelict land for recreational purposes, such as walking, cycling or horse-riding. A principal source of land which furnished the basis for such applications, because of its highly linear character, was to be found in disused railways, of which some five and a half thousand miles had been closed in England and Wales alone since the end of the Second World War.

By the beginning of 1969 applications for grant were beginning to arrive at the Countryside Commission for approval or rejection. The Commission, however, had few guide-lines, if any, to enable it to make consistent, rational, and above all fair and sensible decisions likely to be acceptable to the public. It was therefore decided to commission a Report on the whole subject, and, presumably because of my known interest in railway geography, I was asked to undertake this in the summer of that year. Of the very important legal and engineering aspects of the subject I knew next to nothing, but then I didn't suppose that any one person, however expert in those fields, knew about all the other dimensions of the problem, and one thing about geographers is that their training and experience has inevitably encouraged them to handle material from widely disparate sources; so I felt confident I could find out enough to sketch in the broad outlines of the picture, which seemed to be what the Commission wanted, and quickly!

There was just one area where I felt insufficiently competent to evaluate the potential of these lines and that was if they were to revert to agricultural use. Fortunately Richard had just completed his studies at the Lancashire College of Agriculture and had a long vacation to fill in with something useful before embarking on a post-graduate course at Seale Hayne College in Devon. We therefore formed a father-and-son team and worked like beavers through that summer. Before the end of the year I was able to present our Report to the Commission and it was published in the following year (Appleton, 1970).

Although I suppose it could be argued that disused railways had something to do with transport, this was, in fact, more like a land-use study, and it seemed to have jolted me out of the rut in which my previous work was in danger of getting stuck. I now had no work 'on the stocks' and my mind began to turn towards an idea which I had entertained for some time, namely to argue in a series of essays for a recognition of the importance of landscape in geography. I can't now remember what all the essay titles were to be. Not that it matters, because one of them was destined to outgrow the format of a short essay and, like the young cuckoo in the nest, to heave out all the others. It started from a not-too-serious paper which I had read a few years earlier to a gathering of students in Edinburgh under the title 'The attitude to landscape'. By 1970 I had taken my rough notes out of the drawer where they had been reposing and I started giving more serious thought to two simple questions, 'What do we like in landscape?' and 'Why do we like it?'.

Proclivity and Prejudice

My first point, then, is that the decision to address these particular questions was dependent on a long sequence of accidents, often beyond my control and sometimes involving sudden and unexpected changes of direction. Had any of these events been different the whole process could have been aborted. The second point is that, even when this point had been reached, the nature of my investigating instrument, my brain, was equally a product of the same life-story. If objectivity is regarded as desirable in the pursuit of truth it's important to realise how suspect is the equipment through which we seek to reach it, and this warrants a further look at the instrument in question.

All observers of the environment start with information acquired by the sense-organs. This becomes the input fed into the brain. It's there that we build our own images of the world outside. In the terminology we've been using it's there that we each set about making our own world. When that world has been made it becomes available to be used as a model for whatever purposes we choose, and not least for discharging those tasks which are essential if we are to survive in that other world, the 'real' world, in which we have to live.

Let me then point out just a few things which I think may be relevant to a better understanding of the working of my own neural system, its potentialities, its limitations and the various kinds of bias which may be expected from an instrument which has been subjected to the particular set of vicissitudes described in the previous pages. One of the first things it had to come to terms with in childhood was that the body in which it was lodged was not a very big or a very strong one. Of course not, you will say! All children are small and weak in comparison with adults! But I was small and weak in comparison with other children of my own age. I grew to the average height for an Englishman later, but when I went to the Glebe House at the age of nine I was the smallest, though by no means the youngest boy in the school.

It seems to me highly probably that this must have had a bearing on the fact that I was a cautious and often rather anxious little boy, as must be apparent from my confessions about my absurd neophobia in foods. This lack of self-confidence is also consistent with my concern to inform myself about my physical environment, in case it should be harbouring some source of danger of which I ought to be aware. It could therefore be a contributory factor in the enjoyment which I found in environmental observation and in the precocious development of my sense of orientation, out of which arose all sorts of less obviously consequential quirks, such as devising a mental picture of numbers arranged in the same shape as a part of the road map of Stibbard (Figure 5).

Perhaps a more obvious consequence of my small stature and consequent lack of self-confidence was an exaggerated sense of fairness, to which I have already referred. I don't put it forward as being necessarily a moral virtue; it can equally be a sign of weakness. The strong are able to dispense with it and still come out on top. It's the small and the weak who have a vested interest in supporting the idea of equality, and the price of securing a fair deal for oneself may be to concede a fair deal for everybody else as well. 'Blessed are the meek' is, no doubt, a worthy sentiment, but it's a particularly worthy one for the meek, and I was one of them.

This early obsession with fairness was encouraged to grow through years of experience on the school playing-fields. Perhaps the only aspect of the ethos of the public school which I totally and permanently absorbed was a capacity to accept rules and the authority of an independent agency to enforce them, even when one personally disagreed with a decision, provided, of course, that, as on the playing-field, the rules could be seen to be equally protective of the rights and interests of everybody concerned. This respect for the observing of rules has persisted, and when I see so-called sporting personalities arguing with referees, refusing to accept umpires' decisions and attempting to get their own way by having recourse to tantrums, I find it hard to regard them with anything but contempt, whoever they may be.

A commitment to fairness was therefore an important part of my make-up. It was this which caused me to be so incensed by the chauvinism of the Oxford Union, and it found expression in almost every facet of my life. It has a direct bearing on the way in which the brain weighs one argument against another and it can extend into the field of environmental perception by insisting on the according of proper weight to potentially conflicting phenomena.

One consequence of this is that I'm not a very good campaigner on environmental issues, even those about which I have strong feelings. I tend to argue for the other side of the question in order to keep things in perspective. This is good for the maintenance of fairness, not so good for making firm decisions. Decision, however, is a pre-requisite of commitment, and commitment a pre-requisite of action. If, for any reason, such as a concern to be fair, decisions are deferred, action is inevitably postponed. Whether we call this 'sitting on the fence' or 'keeping an open mind' depends on whether we wish to imply approval or the opposite. Men and women of action are men and women of decision. To such people we look for leadership, for commitment, but not primarily for objective judgement. Society needs the uncommitted as well. In a parliamentary democracy it's the floating voter who chooses the government.

In addition to my small stature I suspect there may have been other characteristics of my physical self which may also have influenced my habits of perception. For example, while I can't prove a connection between long-sightedness and a fascination for very long views, my long-distance vision has always been very good. Had I been naturally short-sighted it seems unlikely that I would have become hooked on a kind of pleasure which I was not able to achieve efficiently.

Apart from physical characteristics of this sort, it's probable that all kinds of other influences have affected the ways in which I've tended to explore the world around me. For example, although I can read reasonably fast when I want to, the *sound* of words is so much a part of the experience of reading that I get little enjoyment from rushing on at a pace faster than that at which I would read aloud. The resonance of the words, the way they build up into phrases, the crucial importance of emphasis, of cadences, of 'dynamics' in the art of elocution, these things are wholly lost if one is content to allow the eye to fly over the pages with no restraint. So, when I'm reading for pleasure I read desperately slowly and get through comparatively little. I must have discovered far less about the world through the printed word than most academics, and in general I've found it more rewarding to make my discoveries more immediately, more quickly and more meaningfully by using my own direct powers of observation in acts of personal exploration than by processing the second-hand accounts of observations by others, however illustrious their authors may be. If this makes me a Philistine, too bad! Where, of course, documentary information of any kind supplements, amplifies, complements or clarifies direct observation that's a different matter. Fitting historical documents, through maps, to landscape as observed in the field (or if necessary through projected slides) was central to my teaching techniques as well as to my own enjoyment of historical geography.

I suspect, also, that this emphasis on direct visual investigation may have a bearing on the much greater ease with which I can handle concrete as compared with abstract ideas. This may just be a sign of intellectual weakness (I'm sure it is), but it's a fact that, whenever I have to express abstract concepts, I'm much more inclined than most people to have frequent recourse to concrete examples, metaphors, similes, parables and the like, as you may have noticed, and I have great difficulty in following the abstract expressions of others when they do not. Often it's bad enough when they do!

Just as I suspect that my keen sense of geographical orientation must have been encouraged by those real-life exploratory experiences, so I believe my visual imagination may have benefited rather than suffered from the fact that the input of fiction came to me as much through the ear as through the eye. As a training for translating verbal into visual images there must have been advantages in being brought up in the pre-television age. Apart from the stories told to us by my mother, supported, perhaps, by a rather sparse back-up of illustrations, our most regular source of fiction input came from the Children's Hour of the B. B. C. with no illustrative back-up at all. Here was a kind of alternative society, a parallel world populated by honorary Uncles and Aunties which contained an immense store of raw material for the fashioning of visual images. From this repository items were selected by one of the said Uncles or Aunties and transmitted through the ether to emerge, at first through headphones and later loudspeaker, in the cosy atmosphere of the pre-bedtime fireside. The messages arrived in code, and we had the daily practice of decoding them from the verbal to the visual form.

So the training of the physical eye to apprehend the detail of the real world went hand in hand with the training of the mind's eye to penetrate the world of literature in pursuit of its own self-painted pictures. 'How true!' I thought to myself when I heard Alistair Cooke, that great draughtsman of verbal pictures of North America, recount the story of the little boy who preferred sound radio to television 'because the pictures are better'!

Some people dismiss the idea that their adult habits of thought and behaviour are deeply influenced by their education. They generally confuse habits of thought with opinions on issues, and I doubt whether anyone can honestly say that the impact of several years of educational experience can ever be completely obliterated. The process of rejecting opinions which had been encouraged if not inculcated by teachers in educational institutions was well under way before I left Shrewsbury. I had begun to realise, for example, that the rules by which most societies (including schools) govern their own members are *not* equally protective of the rights and interests of all. A dedication to fairness might then logically lead to a questioning of the rules themselves. So I found myself breaking away from much of the dogma which had underlain the process of my upbringing and extending this revisionism into politics, philosophy and religion.

In politics I became an Independent ('Sitting on the fence?') but with a commitment to the idea of proportional representation, again on grounds of 'fairness'. I still find it unacceptable that practically every British Government in my lifetime has assumed office after an election in which a substantial majority of voters had indicated that they didn't want it, so I've consistently voted for parties which have supported electoral reform, but I've never joined a party. In religion I never rejected the Christianity in which I had been brought up as a 'son of the Manse'; I recognise it as the principal source of my ethical code, such as it is, but increasingly I've come to see it as metaphor, seeking to penetrate, like poetry, those areas of human experience which lie beyond the reach of reason. I find a lot of its theology no longer meaningful if interpreted literally, while Fundamentalism of every sort seems to me intellectually indefensible and often downright mischievous.

I don't wish to pursue either politics or religion further in the present context, but the third member of the trilogy, philosophy, perhaps has a more immediate relevance. The only philosophy which I knew anything about when I left school was the philosophy of the ancient world which I had encountered to some extent in my studies of the Classics. I began to realise, for example, that, whereas the public school ethos was a product of Stoicism, and whereas I had been taught to regard the doctrines of Epicurus as false, perhaps the time had come to weigh the evidence for myself and come to my own opinion.

The Stoics had argued that it was incumbent on every individual to discover what was 'right' and to pursue it irrespective of the consequences. Emphasis on the rejection of passion and desire led to austerity. Personal inconvenience or distress must not be allowed to stand in the way of what was conceived as moral behaviour, hence the proverbial 'stiff upper lip'. This philosophy had been opposed by Epicurus who had argued among other things for the rejection of the fear of superstition and of death and for the setting up of objectives which lay within the reach of the individual's attainment rather than the determination of moral goals which might well lie beyond it. If, therefore, there was such a thing as absolute right, as the Stoic philosophy implied, whether revealed by the gods or by oracles or by conscience or by some other authority, Epicureanism seemed to involve a betrayal of the principal that its claims were absolute.

Under centuries of Christianity, which preached a system of morality based on the concept of absolute truth as revealed by the Old and New Testament rather than by the gods, oracles, soothsayers and philosophers of ancient Greece, the principle of unswerving commitment implicit in Stoicism was sacrosanct and 'epicure' became a dirty word. The philosophy itself was caricatured as a system of belief which allowed one to do anything one liked in pursuit of unbridled pleasure. Although this is, in fact, almost the opposite of what Epicurus taught, it was an interpretation which prospered in Western Europe through centuries of religious bigotry. The more puritanical the régime the more rabidly did it represent

Epicureanism as the embodiment of immorality. It came as something of a shock, therefore, to discover what Epicurus actually said, and even more of a shock to realise that I was at heart no Stoic but, if I had to come down off the fence, an Epicurean!

The same determination to reject the seduction of the package-deal (described in Chapter 3) and to build up independently my own sense of values was applied to the arts. The infallibility of the arbiters of 'good taste', what I was later to call 'the aesthetic priesthood' (Appleton 1990), seemed at least as indefensible as that of the rule-makers in politics, philosophy and religion. Although the re-appraisal of values in the arts took many forms and involved many kinds of artistic expression, a central theme was the apparent conflict between the Classical, in which I had been basically educated, and the Romantic, which I began to realise, as I grew older, in some ways made the more powerful impact.

I doubt whether anyone has ever defined these two terms satisfactorily, and the more familiar one becomes with the work of artists in every medium the more does one become aware of the difficulty of doing so. But I suppose there would be some agreement that a classical author, painter, architect or composer goes quite a long way in accepting limitations purposefully imposed on his freedom of action. In all the classical arts there are constraints which actually contribute to and enhance the value of the work. It would be easier to do a round of golf in par if the holes were made bigger, but this wouldn't necessarily make it a better game; and by the same token, easing the 'rules' of any classical art form may permit one to achieve what would otherwise be impossible, but always at a price.

I don't think my acceptance of the restrains of classicism was due simply to a feeling of obligation, a sense that what my teachers praised must be good. I've already admitted to an early liking for Fitzgerald's translation of *Omar Khayyám* and for the poetry of Housman, both of whom subjected themselves to the strictest demands of metre, rhyme and style. When I have dabbled in the writing of poetry myself I have occasionally tried free verse, but no sooner have I begun than I find phrases forming themselves in conformity with some metrical pattern and it isn't long before I'm into metrical verse (Appleton, 1978).

The same goes for my taste in music. When it was first awakened it showed a strong preference for the German classics (Chapter 6), though the Romantic was never absent from the repertory which I enjoyed. Later, Romanticism began to play a more important role, but even then simplicity, rigid rules of harmony and a singable tune remained, for me, the hallmarks of excellence, and I think it would be perverse to deny that these preferences probably had something to do with a grounding in the Classics and the resulting habits of style from which I was slowly and only partially to emerge.

On the whole I've found the abandonment of strict rules easier to come to terms with in poetry than in music; so, while I can't *write* free verse with any sense of conviction, that doesn't mean I can't find enjoyment in some of the work of those who can. I'll freely confess, however, that it's been with a sense of satisfaction that I find many present-day poets, and not least my late friend and colleague at Hull University, Philip Larkin, reverting to a fairly strict framework of formality. In Philip's case this also applied to his taste in jazz, on which he was a leading authority and a regular writer. It was the old rhythmic, melodic, traditional jazz and its derivatives that he found exciting. When it began to break up in an apparently random search for novelty, untrammelled by the discipline of pulsating rhythm and the same basic rules of harmony which would have been familiar to J. S. Bach, he lost interest. So did I.

Underlying these preferences there seems to be a dichotomy, which I think is to be found in all the arts, between, on the one side, order, regularity, and harmony, and, on the other, disorder, irregularity, and discord. Only in the loosest way can classicism be equated with the former and Romanticism with the latter. Both involve themselves deeply in both camps;. but Romanticism asserts its right, as it were, to break out somewhat further, and with assurance, from the limitations of the former, far enough, in fact, to achieve new objectives without entirely forfeiting the rewards which derive from a self-imposed subjugation to that formalism and that sense of underlying order which Classicism cannot forsake without forfeiting what makes it classical.

Many of my favourite works of art, particularly in music, now lie on the Romantic side of the divide, if divide there be. I could still listen indefinitely to Vivaldi, Bach and Mozart, but also to Berlioz, Smetana and Dvořák, Bruckner and Mahler. If I had to choose a favourite composer, an exercise which would go very much against the grain, I think it would have to be Schubert. For me the really important dividing-line doesn't concern such things as period or nationality, much less the rankings in the pundits' tables of excellence; it comes between those works which have 'meaning' *for me* and those which have not. By 'meaning' I don't have in mind a rational, logical meaning. In such programme music as I enjoy, for example, the programme is often unimportant and I may well ignore it altogether and build my own pictorial images out of the sounds. When, however, the music breaks so far away from the basic foundations to which I've become accustomed that it loses the structural capacity to support the sort of images (built out of melody and rhythm, harmony and counterpoint) to which I can relate, I easily give up. I definitely prefer my Webern without the 'n'. In a great deal of modern music I won't say the potential to arouse me isn't there, but I don't have the technique to find it.

When, therefore, I began seriously to ask myself how my tastes in landscape came to be as they are, I could find some broad hints by looking in the mirror. There I saw an information-processing organism with numerous design faults of bias, prejudice and a capacity for irrational selective emphasis; and if these idiosyncrasies had any bearing on the way my tastes had developed in poetry, music and the other arts, must they not also have affected the way I look at landscape?

If, for example, I were to compare a park laid out by Capability Brown with one of the great seventeenth-century set-pieces of André le Nôtre or his followers, or with the near-wilderness exploration-grounds of the Picturesque, like Hawkstone, it became clear that the dichotomy between order and regularity on the one hand, and disorder and irregularity on the other, although of immense importance, could not *consistently* explain my preferences. Perhaps my predilection for the happy medium would promote the Brown design to the place of honour; perhaps my classical antecedents would favour Versailles; perhaps my sensitivity to the world of nature would elbow out both of them and leave the pine-and-precipice formula of Hawkstone as the winner.

An understanding of my attitude to other kinds of experience might lead you to expect me to be reluctant to make a choice, and you'd be right. Each has its own virtues, each its own charm. If we could say that the more orderly a design might be (or, for that matter, the more disorderly) the more it would please, or if we could see that a perfect balance between order and disorder invariably made for excellence, we would have established some general principle from which a rational system of preferences could be established. But if preference is not consistent, if different people seem to like different things, or if the same person likes one thing at one moment and another thing at another, we may well come to the conclusion that we're left with a catalogue of disparate items not recognisably connected by any principle at all.

Nor are we any better off if we apply the same approach to ordinary non-designed, uncontrived landscapes. Sometimes I find the heathlands, with their pines and silver birches sporadically rising from a sea of heather, the most attractive type of landscape I can think of, but a short period of exposure on the bare mountains or of deep seclusion in the oak forest might well establish each of these, for a time, as the most desirable landscape type. Exposure and enclosure are clearly of crucial importance and can totally alter the feel of one's surroundings, but as absolute criteria of aesthetic values they would be no more valid than order and chaos, civilisation and wilderness, familiarity and novelty, simplicity and complexity or any other pair of presumed opposites.

In short, while the ancient Greek proverb 'Know thyself' could make an important contribution to one's own interpretation of one's own landscape tastes in terms of one's own psychological make-up and one's own past experiences, that contribution would still fall far short of providing a key to any universal laws. It's one thing to reject the package-deal and to isolate ideas one can believe in; it's quite another to find a connecting system, a body of principle to tie all those isolated ideas together. Round about 1970, when I was fifty, I began to look more searchingly for such a linking principle, which, if it existed, would make more sense than a mere recital of a catalogue of likes and dislikes.

9   RERUM COGNOSCERE CAUSAS

'I looked for universal things; perused
The common countenance of earth and sky.'

In one of the best known lines from *The Georgics* Virgil has provided
the motto for all researchers. *Felix qui potuit rerum cognoscere causas*
can be loosely translated as 'Happy is the person who has been able
to discover (recognise, understand) the causes of things', and when I
began expanding my little paper on 'The attitude to landscape' from
a somewhat light-hearted and informal if not facetious essay into a
more serious book I soon realised that, although few geographers
had yet strayed into this area, the urge to understand had already
drawn into it scholars working in a number of different disciplines:
philosophers, psychologists, art critics and historians, literary critics,
conservationists, landscape architects and others with practical as
well as theoretical interests. I therefore spent a good deal of time
expanding my elementary knowledge of several subjects with a view
to finding out how far they had progressed. I achieved a smattering
of knowledge about these subjects but not much more.

My natural apprehension at trespassing in other people's fields
had been allayed, when I was working on disused railways, by the
knowledge that, if I knew little about their territory, they probably
knew even less about mine or even each other's. I had long been
accustomed to geography being regarded as a scrap-book of bits and
pieces culled from other disciplines. This was its greatest weakness,
but I soon realised that, in the present context, it could also be its
greatest strength. The academic world is a world of specialists, each
pursuing research along a restricted frontier of knowledge, each
aspiring to be a leading authority on his or her own tiny sector of that
frontier, yet each often fairly ignorant of even quite elementary
concepts in other people's specialisms.

What was needed was a capacity for 'lateral thinking', as it came
to be called, and my training for this could have been much worse.

After all, I had at least dabbled in such diverse subjects as the Classics and geology, and I had learnt how to retrieve information from the literature of history, economics, sociology, anthropology and many other subjects for the purpose of applying a wide range of data to geographical ends. Jack of all trades I might be, but so what? I was able to acquire from these reflections enough confidence to trespass without too many inhibitions.

Not unreasonably I imagined when I began, that the answers to all the questions I had been asking must already have been discovered by philosophers. After all, they had been pursuing them for at least a couple of thousand years, so surely this must all have been settled long ago. I had touched on the work of the ancient philosophers at school and had encountered Plato among my set books in 'Pass Mods' at Oxford, but the emphasis had been on what used to be called metaphysics and ethics; I had virtually no knowledge of the philosophy of aesthetics, so I turned to this literature with eager expectations.

I'm bound to confess that, in general, I found these works disappointing for my purposes. Like much philosophy the literature of aesthetics proved to be largely a literature of abstract language, and I've already admitted that my thinking apparatus was geared to handle raw data more easily in a concrete form. Between the world of hills and valleys, summer meadows and Scots pines, quiet gardens and vast panoramic views on the one hand and concepts like proportion, expression and intuition on the other there was a great gulf fixed. The world I had made belonged to the former; the philosophers seemed to be wholly immersed in the latter, and whether it was my fault or theirs, I was forced to conclude that, for me at least, this wasn't the right starting-point. So, instead of starting with abstract concepts, I asked myself what would happen if one began by looking at the components of landscape themselves. Perhaps one could find in mountains, lakes, woods or different types of buildings inherent qualities which could explain the apparent aesthetic values of the landscapes of which they were the components.

I soon discovered that this approach also was likely to be of only limited value. If the same objects were liked by some but not by others it was difficult to see how any of their physical properties could determine whether they were intrinsically 'good' or 'bad'. I remembered enough of my Greek to recollect that the word 'aesthetic' is an adjective deriving from the ordinary verb meaning 'to perceive'. The connotation of beauty became added later, first by infiltration, then by take-over, but aesthetic experience is still literally the experience of perception, and perception is the activity through which individuals inform themselves of their surroundings. As we have seen, the acquisition of environmental information is a pre-requisite of environmental adaptation, which in turn lies at the heart of Darwinian evolutionary theory.

Darwinism and Aesthetics

There seem with hindsight to be several reasons why I chose environmental perception in a Darwinian context as a starting-point. One was the status which I had accorded to Darwin as a kind of schoolboy hero; long before I had read *The Origin of Species* I had felt myself to be emotionally on his side. Another was to place more weight on my own observations, as compared with the transmitted wisdom of others, than would most people. Yet another was my scepticism for the package-deal.

It could, of course, be argued that, in turning to Darwinism, I might be said to be accepting just such a package; but my objection was to the wholesale, uncritical acceptance of the package, whether it were Stoicism, Judaism, Christianity, Islam, Marxism or anything else, not the selective acceptance of its various contents. And whereas most religious, philosophical or political dogma contains among its own tenets strict injunctions against revisionism, the teaching of Darwin included built-in assumptions that such revision would be a constant, continuing and absolutely necessary part of the investigative process to which he saw himself as contributing a central but certainly not a final statement. *The Origin of Species* was published with much reluctance and was seen by Darwin himself as a prematurely enforced *interim* contribution to an on-going debate. The last thing he would have claimed for it was that it should be seen as the definitive and authoritative revelation of the whole truth, and we are now in a position to see how right he was. Many of the minutiae of his thesis have indeed been revised, and many of the gaps, of which he was fully conscious, have been filled; but his reputation, except in the eyes of the scriptural fundamentalists, has survived.

What particularly appealed to me about Darwinism in the present context was that, by its own terms of reference, it had to comprehend the whole living world. This turned my attention to the arguments put forward by Robert Ardrey and Desmond Morris that we can learn much about the behaviour of our own species by observing that of the other creatures who inhabit the earth. Not only that, but the theme of environmental adaptation central to Darwinian theory assured a major role for environmental perception and for the study of the relationship between creatures (including human beings) and their habitats. Was it possible, I wondered, to build an aesthetic theory starting from these premises?

Every individual is born with certain tendencies, call them instincts if you like, to behave in ways which are conducive to its survival. If it doesn't possess those tendencies its chances of survival are correspondingly diminished. It's the tendencies of the survivors which, by the process of selection, become the characteristics of successive generations of the species. Yet the tendencies themselves are not sufficiently clear-cut to determine the detailed responses of those individuals to environmental challenges. Only when they have been refined by experience will they provide the organism with the techniques necessary to preserve its own interests. Every kind of activity - eating, drinking, self-defending, mating, being protective towards its young until they are old enough to fend for themselves - can be seen to be made up of these two components, the inborn or innate and the acquired or learned.

All this is commonplace to any biologist, and most would readily accept that it's equally applicable to *Homo sapiens* as to any other species. To some non-biologists, however, this is anathema. The human brain, they will argue, is a different kind of brain from that of all other animals, and in this, of course, they're right. But we mustn't think of the animal brain as having been replaced in our own species by a brain which is totally different in all respects. Something very like the simple limbic brain of quite primitive animals is to be found also in mammals but functioning in association with a much more complex neo-cortex. In the human species the neo-cortex itself is much more complicated still and capable of vastly more advanced achievements even than those of other mammals, but we still have a vestigial limbic brain as well, not unlike that of other mammals, and it's there that we experience primitive feelings like pleasure, anger, fear, etc. (Campbell, 1973).

This isn't the place to set out the evidence for this view. The unconvinced may make reference to an already extensive literature. The argument is that the same kind of mechanism, which leads other animals to frequent places in which their chances of survival are high, also operates in ourselves, and achieves its objective by reliance on the development, by a combination of heredity and experience, of some system of preference. Those who, for whatever reason, find themselves unable to accept this argument may well have difficulty also in accepting the arguments which follow. Suffice it to say that they would be out of step with the great majority of biologists, neurophysiologists, psychologists and behavioural scientists, and that, if we are in dispute about this, we must refer it to the judgment of posterity.

The model, then, of a creature of whatever species embarking on the task of surviving in a world which offers opportunities for both success and disaster is something like this. It enters the world 'programmed', as it were, to seize the advantages afforded by the environment while avoiding its disadvantages. Many of the simpler physical functions it needs to perform don't need to be learned by experience. It's immediately able to breathe, for example, as well as to operate those processes, like the beating of the heart, over which it has no voluntary control anyway. Even some of its defensive reactions are already a part of its natural instinctive behaviour, like crying for help, but others will need to be developed, sharpened and made more sensitive to particular signs of danger, which it must learn to recognise, through play, for example, or in the course of its defence mechanisms being employed in earnest.

Actions of this kind are indispensable; they *must* be put into action if the creature is to survive, and this means that there *must* be some mechanism which ensures that they are. That mechanism is what we call, for want of a better word, 'pleasure'. There are plenty of other words like 'desire', 'drive' or 'libido' which one may find employed in the literature. In plain language we do all these things on which our survival depends *because we want to*. That is the force which impels us.

Among the various activities which are essential to our survival there is one which, in our present enquiry, is crucial. If, as I've suggested, environmental adaptation is dependent on environmental perception, then it follows that this too must be put into regular practice; and the same mechanism must operate to ensure that it happens. The systematic perception of unfamiliar environments is what we call exploration, and we need to be motivated to practise it relentlessly by that same driving force, the pursuit of that pleasure or satisfaction which is attained in the very act of performing it.

If at this point we pause to remind ourselves that our purpose is to trace the steps by which one individual acquires particular tastes and preferences in landscape, we can see that the line of reasoning we have just pursued is beginning to take on an obvious significance. We have, in short, injected the pleasure principle into the aesthetic process. We can summarise the argument somehow like this. The achievement of physical survival requires the acquisition of adequate environmental information. This is provided through efficient environmental perception, the motivation for which is in turn provided by an innate desire to explore our surroundings, and the gratification of that desire is the feeling of pleasure which accompanies its performance (Kaplan and Kaplan, 1982). We enjoy it for its own sake. If landscape may be defined as 'the visually perceived environment', landscape taste is 'a system of preferences for particular kinds of perceived environments', while those preferences, like all preferences, may be seen to be made up partly of innate and partly of learned components. It's this last point which

leads us to expect that our own individual experiences are likely to
have played a substantial formative role in the development of
unique individual ways of looking at landscape and of distinctive
idiosyncratic preferences for what we see in it.

When, some time about nineteen seventy, I found myself
beginning to think more systematically along these lines, I
withdrew from my storehouse of recollections the little incident of
Pembroke-im-Pitztal (p.133). It was then that I began to see a new
significance in a trivial and fleeting day-dream. The central notion
which had persisted in the fusion of the two images (of Pembroke
Castle and the Austrian valley) was that, in spite of the great
differences of scale, they both induced a common sensation, that of
being in an enclosed haven of security beyond the rim of which lay
the wide world outside. The fact that I personally couldn't see over
the margin with the physical sense of sight seemed to attach more
rather than less significance to those limits imposed on my powers
of vision by the encircling screens. They became critical boundaries
between that part of the world which I could see and that part
which I could not, the field of perception and the field of
speculation respectively, the domain of the eye and the domain of
the imagination.

At about the same time that these thoughts were passing
through my mind another trivial incident occurred which played an
important part in their future development. Some friends of ours,
Guy and Heather Greenwood, who lived in the Lake District, went
on holiday and invited Iris and me to spend a few days walking in
the hills and at the same time keeping an eye on their house in their
absence. One night, before going to sleep, I picked up a copy of *King
Solomon's Ring* by the biologist Konrad Lorenz, which my host and
hostess had left by my bedside. I felt by this time that I had the
broad outline of an argument which needed some kind of focal point
to bring it together, and, while thumbing over the pages of this book I
stumbled on the following passage;-

> It is early one morning at the beginning of March,
> when Easter is already in the air, and we are taking a
> walk in the forest whose wooded slopes of tall beeches
> can be equalled in beauty by few and surpassed by
> none. We approach a forest glade. The tall smooth
> trunks of the beeches give place to the hornbeam
> which are dotted from top to bottom with pale green
> foliage. We now tread slowly and more carefully.
> Before we break through the last bushes and out of
> cover on to the free expanse of the meadow, we do
> what all wild animals and all good naturalists, wild

boars, leopards, hunters and zoologists would do
under similar circumstances: we reconnoitre, seeking,
before we leave our cover, to gain from it the
advantage which it can offer alike to hunter and
hunted - namely to see without being seen. (Lorenz,
1964 edn., p.181).

It very quickly dawned on me that the last few lines of this
passage contained at least three ideas which were of immediate
relevance to the solution of the problem of bringing so many loosely
connected arguments into some sort of system of environmental
aesthetics, though it was only later that I realised how important they
were to become. The first thing that excited me was to find one of
the world's leading behavioural scientists incorporating people and
(other) animals in a single category of creatures whose behaviour, at
least in one particular respect, could be described without
discriminating between naturalists and wild boars, leopards and
zoologists. This confirmed my growing opinion that, at a
fundamental level of behaviour, we all respond to environmental
stimuli in similar ways, and that it should be possible, *given the
necessary safeguards*, to draw inferences about our own behaviour
from that of animals, as Ardrey and Morris had so powerfully
argued.

The second idea was that the necessity to make a careful
appraisal of the environment is binding on all creatures whether they
are seeking food as predators or endeavouring to avoid becoming
some other creature's prey. In other words the same basic rules
apply irrespective of the roles of the observers. Hunter and hunted,
human and non-human, have the same basic needs for
environmental information.

It was, however, the third idea which, as it were, broke the log-
jam and opened up the way ahead. 'To see without being seen'
seemed to encapsulate the essence of the whole exercise (Figure 47).
Exploration is essentially a strategic operation, calculated to secure
an environmental advantage, and two of the most critical
components of the exploring situation are seeing and hiding.

Prospect, Refuge and Habitat

From this simple starting-point I went on to develop a typology of
landscape based initially on these two criteria, how far a place was
able to afford opportunities, on the one hand, to see and, on the
other, to hide or shelter, applying the words 'prospect' and 'refuge' to
each concept respectively. This became the central theme of the book
which had grown out of my little essay on the attitude to landscape

FIGURE 47. 'TO SEE WITHOUT BEING SEEN'. A fox cub demonstrates the symbolism of prospect and refuge. Photo by Richard A. Brown taken on the Bloedel Reserve, Bainbridge Island, Washington, U.S.A.

FIGURE 48. 'WATCH ME, MUMMY!' Charles Appleton jumps into a river at the Pine Forest, Armidale, New South Wales, displaying an abnormally high threshold of hazard tolerance for a four-year-old. Photo by the author, 1961.

and which was published under the title *The Experience of Landscape* in 1975.

Its publication was the most important step in the progress of my world-making for two reasons. In the first place it was my first attempt to set down on paper something like an explanatory model of landscape aesthetics. It was an *interim* statement, crude, reductionist and over-simplified no doubt, but it had the effect of injecting into an ongoing debate certain hypotheses which, as I explained in the previous chapter, had arisen out of my own personal experiences every bit as much as the idiosyncrasies of my personal tastes in landscape.

In the second place, as these ideas gradually became familiar to other people who were writing on landscape, and as I in turn became acquainted with their work, I began to feel myself to be a member of a larger movement, almost, one might say an evolutionary school of landscape aesthetics, distasteful as it might be to find myself involved, however marginally, in putting together a 'package'. Perhaps it's too pretentious to speak of a movement or a school, but at the very least I found myself forming acquaintances, even friendships, with people who were also actively seeking answers to questions similar to those I had been asking myself. One important consequence was that, from this time onwards, I was even more exposed to the danger of assimilating ideas from other people without being fully aware that I was doing so. Plagiarism, the deliberate representation of the work of other authors as if it were one's own, must rightly be condemned, but we can never be quite sure how far what we presume to be our own ideas may include the products of other people's thought-processes, and this danger is increased when one begins to exchange ideas through unrecorded conversations with other people who share one's interests.

Who, then, were these people with whom I was beginning to exchange ideas and did they have any common characteristics which might justify the appellation of a movement or a school? First, although they are mostly professionally involved in academic institutions, they cover a wide range of disciplines - geography, biology and its various sub-divisions, psychology, architecture, landscape design, literature, art history and several others. This enables them to contribute expertise from their own specialisms and to apply it to a much wider field.

Secondly they are in some degree sympathetic to the idea that, before we designate 'aesthetic values' as belonging to some alternative system of values, having nothing to do with the values of the material world, we should first examine closely the possibility that they may be the products, at least in part, of those same laws of the biological sciences which explain other facets of our behaviour as animal organisms.

This means, thirdly, that they subscribe to the view that the various forms of living creatures and their associated patterns of behaviour have been the product of some sort of evolutionary process. Most of them, therefore, would attach importance to such phenomena as environmental adaptation, the relative influence of heredity and experience on behaviour, the behaviour of creatures other than *Homo sapiens*, and the neural processes by which environmental perception is achieved. All of these views would be likely to bring them into conflict with some of the more conservative and traditional aestheticians, who still seem to be deeply suspicious of any reference to the biological basis of human behaviour, and regard it as being at best only marginally relevant to aesthetics and at worst downright mischievous.

The proposition to which I found myself drawn in the early seventies, in short, envisaged our own species as perceiving landscape in much the same way as other creatures perceive their habitats. The concept of habitat implies that every species has its own set of needs peculiar to itself, some of which it shares with other species and some of which it does not, that different places are endowed in varying degrees with a capacity to satisfy those needs, and that each individual is born with a natural inclination to frequent the kind of habitat appropriate to its species. The most favourable places within such a habitat are likely to attract the largest number of individuals, and this may be perceived in two contrasting ways. On the one hand it may suggest that it has already been judged by others to be a suitable place, and this will further enhance its perceived value. On the other hand, it may be seen as implying competition for the benefits the place can afford, as foreshadowing, in consequence, a struggle for the right to occupy it, and as indicating the possibility of ultimate failure to realise its promised potential. A balance must therefore be struck between its perceived advantages and disadvantages. If the latter are deemed to outweigh the former the site will be rejected and another will be sought. This is the basis of habitat selection (Orians, 1980).

A habitat, for most species, must meet two complementary sets of needs. The first is an adequate supply of food, including water; the second is a particular place, within or accessible to the former, which will prove suitable for raising a family of young in comparative safety. These are termed 'foraging-grounds' and 'nesting-places' respectively. Many other considerations enter into the determination of whether a habitat is suitable, not least the probability of being able to carry out these operations without interference by predators.

Among some species, individuals or mating pairs are prepared to share foraging-grounds with others of the same species, and there may be some advantages in doing this on the principle that there is safety in numbers; but in other species individuals or pairs are not prepared to do this and will contest their right to an exclusive use of such foraging-grounds. 'Territoriality' can lead to highly aggressive behaviour. It's widespread throughout the animal kingdom and is one of those behavioural characteristics which can most easily be seen to be shared by our own species.

Habitat selection involves the activity of exploration, that is to say organised or systematic environmental perception. An exploring individual seeking a habitat will examine the general lie of the land and, within the wider field, will look for one special place which might prove suitable as a nesting site. When such a site has been chosen the process of habitat selection is complete, but the process of environmental perception continues, because on it depends the efficient use of all the opportunities latent in the chosen site.

It's precisely this kind of argument which so infuriates those who wish to deny that students of human behaviour can learn anything useful from the behaviour of other species. We don't live in nests, they will say, neither are we dependent on foraging in the great out-of-doors. We live in houses or apartments, they will say, and we obtain our food from the supermarket, and of course they're right. But this misses the point. I've already argued that all behaviour-patterns are based partly on innate tendencies or instincts, but that these become modified, refined and channelled along particular lines by learning. Learning in turn may involve an element of teaching, that is to say a deliberate attempt by adults of the species to pass on their own acquired behavioural tendencies, the fruits of their own experiences, to their young.

There's really no logical ground for supposing that this doesn't apply equally to human beings as to other species. Indeed nobody who has participated in the process of bringing up young children could reasonably deny that it does. The main differences which distinguish *Homo sapiens* from other species in this respect are, first, the more complex structure of the brain, already referred to; secondly, the particular demand of the human infant, which remains incapable of looking after itself for several years; thirdly, and largely for that reason, the minimalising of the importance of a seasonal breeding-cycle; fourthly, the highly artificial environment into which the young are born, and, fifthly, the more complicated and efficient machinery for transmitting learned information through a far more sophisticated type of language.

Even if my opponents concede this, however, they will fall back on their next line of defence. This may have happened, they will say, to 'primitive man', but these innate tendencies, if they ever existed, would have died out by now as civilisation has replaced a natural environment by a largely artificial one, just as unused limbs eventually disappear from a species which no longer has need of them. What they overlook is the extremely short time-span, in evolutionary terms, which has elapsed since our ancestors *were* living in conditions not greatly different from those of other primates. Taking a generation as, say, twenty-five years on average, a hundred generations would take us back almost to the beginning of the Iron Age even in the most advanced civilisations. Two hundred would take us back, in a country like Britain, well beyond the beginning of the Bronze Age and into the Neolithic. And while a very few generations may suffice to bring about great changes in physical features by controlled selective breeding, a hundred is nothing in a non-selective evolutionary process.

When, therefore, I came to write *The Experience of Landscape*, the idea that we might still have some kind of inherited ground-rules similar to those which influence the behavioural responses of other creatures towards their habitats, led me to propose what I called 'habitat theory' as a basis for examining the aesthetics of landscape. Having already encountered the antagonism which such an idea was likely to provoke I took the trouble to read what a number of ecologists had written and, having failed to find anything which suggested that what I was proposing conflicted in any way with present-day thinking in that field, I found the encouragement I needed to press on.

The argument, then, is that a human infant at birth has inherited through its genes all the behavioural tendencies essential to its survival. It already wants to breathe, to feed, to be warm (but not too warm) and to be free from those discomforts which are the basis of its warning-system; and it contains the germ of other desires, for example of a sexual kind, which will develop later in its life-cycle, but it will immediately start modifying these innate tendencies by experience, developing an idiosyncratic pattern of preferences which in general conform to the pattern of the species but in detail will be unique to itself, and within this pattern the acquisition and processing of environmental information will be central to its expectation of survival.

Case-study and Theory

Let's now have a closer look at this whole phenomenon of environmental adaptation as manifested in various kinds of desire related to the provision of various kinds of need, glancing back now and again to see how the biographical details of one individual, as recounted in the earlier pages of this book, may fit into the general behaviour pattern of the species. Don't forget, however, that I'm not arguing that these desires are exactly the same as those of other creatures which are attempting to survive under natural conditions; merely that they have evolved from the same biological antecedents. The basic desires to explore a habitat and to establish some degree of user's rights within it are what is common to us all. Once that desire has resulted in a particular process of exploration in a particular place, a particular pattern of familiarity begins to be built up, and this can entirely transform the way we see the place.

The very concept of habitat, a place perceived to be favourable for survival, is implicit in my whole attitude to Stibbard (Chapter 1) and Bradley Hall (Chapter 2). The idea of nesting-place and foraging-ground is embodied in that of the castle with its baileys and the surrounding terrain as a field for exploration, as also in that of indoors and outdoors, usually so clearly defined, but enjoyably unresolved in my grandmother's conservatory; and there could be no more eloquent expression of territoriality than a little boy and his sister identifying themselves with their own 'patch'.

Site selection is not, of course, the only activity to depend on an efficient machinery of environmental perception. Every creature must be able to pick up signals indicating anything which it needs to know about, especially signs of potential danger. Information about sources of food and water, sheltering-places and assembly-grounds may be vitally important, but rarely if ever is it required with such immediacy as information relating to some external threat or hazard. The need for food or water may require to be satisfied within, perhaps, days or even hours, whereas the achievement of security from a potential threat may demand *instant* reaction. It's not surprising, therefore, that information relating to danger needs to be pigeon-holed in a form in which it can be instantly retrieved, and any signal to the brain which is suggestive of possible danger must be able to break in on the consciousness and take precedence over any other mental activity which may be in progress.

In order to ensure that these danger-signs are noticed at the earliest possible moment it's essential that every individual must be perpetually motivated to look for them by the same sort of pleasure-seeking incentive which motivates all other kinds of survival behaviour. If we were to be interested only in those features of our environment which are suggestive of safety, cosiness and comfort, and not at all concerned with those which suggest danger, what sort of a recipe for survival would that be? Seeking the assurance that we can handle danger by actually experiencing it is therefore itself a source of pleasure. This is what the philosophers of two centuries ago meant by 'the sublime'; it's the capacity to induce an acceptable level of awe, even terror, while stopping short of threatening actual danger. Too little exposure to it will not satisfy; too much will induce unacceptable levels of fear, and the pleasure will be destroyed. But what is too little and too much?

I earlier mentioned Jane Gear as one of my painting teachers, (p. 114). She later took a Ph.D., and I found myself sharing in supervising her work which later led to the publication of her book, *Perception and the Evolution of Style: a New Model of Mind* (1989). This is essentially a unifying psychological theory based on environmental perception, and in it she explored further the phenomenon of human variability in the tolerance of 'risk'. Some people seem to need the stimulus of a really dangerous experience before they're satisfied; for others a stroll in the park is an intolerably traumatic event. This led Gear to come up with the concept of *thresholds of hazard tolerance* which went further than I had been able to in 1975 in clarifying the problem of reconciling pleasure and fear.

Edmund Burke, the Irish politician and philosopher, whose *Philosophical Enquiry into the Origins of our Ideas of the Sublime and the Beautiful* (1757) was one of the most influential essays on eighteenth-century landscape aesthetics, wrote:

> When danger or pain press too nearly, they are incapable of giving any delight, and are simply terrible; but at certain distances, and with certain modifications, they may be, and they are delightful, as we every day experience. (Burke, 1958 edition, p.40).

The phrases 'too nearly' and 'at certain distances' are clearly not capable of being quantified in universal laws because, as Gear has shown, each of us has a different threshold of hazard tolerance (Figure 48). Mine comes somewhere near the middle of the range. I've never had any desire to dangle from a precipice or a parachute any more than to penetrate the deeper recesses of a cave. Those ghastly fairground machines which are designed to induce nausea as well as fear hold no attractions for me, and, as I have confessed, my spirit of adventure quickly fails me when I'm confronted by the challenge of unfamiliar food. On the other hand I have many friends who are far more easily daunted than I am by rough terrain or by the fear of losing their way.

If you cast your mind back to some of the incidents I described in Parts I and II you'll easily recognise that the pleasure I found in many of them makes sense when they're seen as exploratory experiments in the assessment of danger. The storm in the Isle of Man and the thunder of Niagara and the Victoria Falls were exciting because they were dangerous; potentially dangerous, that is. I made sure I was near enough to sense that danger but far enough away to be out of its reach.

I have little doubt, either, that it was the symbolism of danger which I found so fascinating in *The King of the Golden River* (p. 47). My mother saw it in the faces of the brothers, inebriated, hideously ugly and brandishing their swords. They lay, if not beyond her threshold of hazard tolerance, certainly beyond what she conceived to be mine, at least at bedtime. I also could see what she saw, but what I remember most vividly is that little thread of golden water plunging from high distant precipices (Figure17), symbol of extreme danger but so far away and so beautifully romanticised that my adventurous little mind was not troubled by it, only compulsively fascinated.

If, then, exploration is concerned with the recognition, as an immediate objective, of the signs of potential danger, it isn't difficult to see it as a strategic exercise which, if it is pleasure-induced, goes some way towards explaining incidents like the games we used to play in those moments of blessed release on the beach at Old Hunstanton (p. 52), not to mention the whole apparatus of military metaphor in the garden at Stibbard with its forts and dug-outs and God-knows-what (p. 13). But I suppose the most cogent experience of that type was in those O.T.C. field days (p. 62) when we were actually required to play at exploiting the advantages of the terrain in a quasi-military context. Had the ammunition been for real it would no doubt have passed the threshold of hazard tolerance of all but the craziest; but of course we knew it wasn't, and I can honestly say I recollect those occasions as highly pleasurable in an essentially aesthetic sense. We just enjoyed the make-believe battle as it was happening.

Perhaps the most enjoyable thing about those field days was the satisfaction which came from watching a piece of landscape in anticipation of detecting some movement which would give away the enemy's position. Earlier I referred to seeing and hiding as complementary components of good strategic practice, and it was in the context of those field days that these twin concepts could be experienced at perhaps their greatest intensity. 'To see without being seen' is a phrase to be found not only in Konrad Lorenz but also in the manual of infantry training. Seeing is, for sighted people, the most effective agency of environmental perception, the other senses playing an often important but usually subsidiary role, and, as in those other activities essential for survival, the necessary motivation comes simply from the pleasure inherent in its performance.

Seen in this light, the desire to see for its own sake, irrespective of the attainment of any immediate practical objective, now seems to offer a rational and perhaps inescapable explanation of those numerous incidents which have been such a source of pleasure to me that I have felt obliged to tell you about them. The views from the attic windows at Stibbard (p. 10) were among my most memorable recollections of the place, and I've also told you about the eastern prospect from the little gate at Bradley Hall across the railway to Rivington Pike (p. 35). The most memorable moments in the journey to Bradley (Chapter 2) were also those where we crested the escarpments of the Carstone, the Lincolnshire Limestone and the Millstone Grit with their panoramic views across the Fens, the Vale of Trent and the Cheshire Plain respectively, the vantage-points becoming progressively higher and the views more extensive as we went further west. These are only a few of the many examples I have given of views rich in the symbolism of 'prospect'.

It's in this context also that I may remind you of the role of horizons as the further limits of visual fields, seen, for example, in the view of the turnpike from my bedroom window at Stibbard and again of Rivington Pike, and not least in the bare lofty rims of the little Pitztal (p. 133). Prominent objects which rise above horizons, like the poplars at Stibbard (p. 21), are really vertical developments of horizons, as are towers and their symbolic counterparts, cumulus clouds. The very concept of elevation communicates the idea of a wider field of vision.

All these phenomena are expressions of the symbolism of 'prospect' in that they are to do with the ability to see. Prospect, however, also finds expression in other conditions conducive to seeing as well as elevated vantage-points. The view from the highest mountain may be frustrated by cloud, mist or fog, as happened on my first two nocturnal ascents of Snowdon (p. 149). So clarity of the air can be said to have prospect value. I now know that this was the significance of what I still remember about my departure down the Mersey from Liverpool at the age of six (p. 37). What I was not to understand for many years was its meteorological explanation. In England, more often than not, there is some atmospheric interference with the passage of light. The greatest likelihood of unimpeded visibility occurs when a cold front has passed through, with brisk north-westerly winds bringing in cold, clear air from more northerly latitudes. The blue skies and choppy water confirm that pretty certainly these were the conditions which obtained on that day when I set out for the Isle of Man in 1926, and the fact that the sea was rough enough to cause problems to me and the Bovril Girl before we reached Douglas is at least consistent with that speculative hypothesis!

The concept of 'refuge', of hiding, of sheltering, of keeping out of the way, is one of the most fundamental in the symbolism of environmental perception. It finds an extreme expression in the search for the nesting-place. If safety can't be secured, and if, in consequence, the individual organism ceases to function biologically, then all other desires become, for that individual, unattainable. Going to the opera may be culturally more prestigious than simple survival behaviour but its attainment is absolutely dependent on staying alive! Small wonder, therefore, that the symbolism of the refuge should retain a prominent place among my early recollections. Stibbard was full of such symbolism (Chapter 1); the cupboards in the nursery, the dark hiding-places, the fort in the shrubbery, the recesses of the yew hedge, the enclosure of the little lawn between our house and the church, the mine and numerous other nooks and crannies were all powerfully associated with the idea of a protected area. Nothing at Stibbard, however, could compare with the gloomy, cavernous, secret world under the rhododendrons at Bradley (p. 34). For me this remained the supreme image of the refuge symbol, even after I had made the acquaintance of the greater glooms of the sitka plantation and the tropical rain-forest.

I'm sure you'll also understand that this kind of symbolism comes in all shapes and sizes. It was present as much in the self-comforting cocoon which I nightly constructed for myself at the Glebe House (p. 53) as in the sense of protective confinement in the enclosing mountains of the Pitztal. Pembroke Castle (p. 132) and Tom Quad at Christ Church (p. 73) exemplified it on an intermediate scale. It was powerfully present in the liturgical shadows of the Russian Orthodox Cathedral in Paris (p. 86) and in the midnight chapel at Amay-sur-Meuse (p. 87).

If the concepts of prospect and refuge are basic to the phenomenon of environmental perception they are merely 'basic'. They are supplemented by many other kinds of symbolism which similarly find expression through objects and arrangements of objects. Often the first reaction to a warning of danger is to lie low. It's then that we need the refuge, and we need it immediately. But lying low, freezing, hoping to escape notice by being inconspicuous, is not the only effective reaction. It may well be that a more advantageous alternative is to get away from the dangerous or uncomfortable place altogether, either immediately or as soon as it seems safe to do so. We therefore have a keen and possibly vital interest in knowing the whereabouts of the most promising escape routes, and these also become the subject of fascination to the exploring eye.

I believe that this idea was powerfully present in those forays which my sister and I used to make along the ditches and under the hedgerows of Stibbard (Figure 7). These were not just places of refuge; they were *channels* of refuge, potential lines of movement along which one could proceed through a landscape between places affording different opportunities which might become more or less beneficial as the strategic situation changed. This again was a concept which found expression at very different scales. It was an escape-route which I saw when I craned out of the bathroom window to look for the trains and to board them emotionally in the imagination as they trundled off to the great wide world of faraway (p. 11). I recognised it in the river valley and in the apparently converging lines of the railway track viewed from innumerable teenage vantage-points (Chapter 6), and not least when I came to enquire officially into the recreational potential of disused railways (Chapter 7).

I've referred to the symbolism of the hiding-place and the escape-route as having a significance of a defensive, protective kind; but *Homo sapiens* is by nature at least in part a predator himself. The species is technically omnivorous in that it takes its food supply from both animal and vegetable substances. 'Hunting and gathering' is a phrase habitually used to describe the provisioning of simple-living communities which rely on collecting their food from whatever nature provides rather than controlling and managing its production themselves.

The acquisition and consumption of food may be seen as two stages in a single process. Normally taste and availability determine what one eats, and in a state of nature a pattern of preference directs the hungry towards a range of foods, in some species very wide, in others very limited, which will provide the nutriment required. Both stages in this process are attended by that essential pleasure without which individuals of whatever species would not be impelled to do what they have to do if they are to survive.

My own conservatism in alimentary taste does not imply that I don't enjoy eating, merely that my enjoyment is confined to a below-average range of preferred foods. Neither does my general aversion to the killing of animals imply that the hunting instinct has withered away, merely that certain forms of hunting are rendered unpleasant for me by the association of my own feelings, however illogically, with those of the hunted animal. So my early interest in fishing - catching fish and subsequently eating them, both of which stages in the feeding process were attended with pleasure - was superseded by a pattern in which the killing part no longer appealed. If I were a wholly logical creature I should no doubt have become a vegetarian, which I have not. That should not surprise you, because by now you will know that logic is far from playing an exclusive role in directing my actions! As for the hunting instinct, I have continued to enjoy the experience of acquisition in all sorts of ways which can be seen as thinly disguised substitutes for the chase.

One such substitute is to go through the motions of stalking a bird or outwitting a fish, but only up to the point at which one can satisfy the desire to establish one's own strategic superiority, stopping short of any action which kills or injures the quarry. So I continued to derive pleasure from watching fish, birds and other animals, preferably in their natural environments, their habitats. But I also found satisfaction in the larger-scale, grown-up equivalents of those acquisitive practices which gave me so much pleasure in childhood - fetching the milk, cockling, and gathering wildflowers and berries. So the harvesting of grain at Wortham Manor (p. 95) and of apples at Bramfield (p. 99) were productive of deep satisfaction, even though the commodities collected were for the benefit of others and not just for me.

I suppose the most unusual examples of a substitute quarry were to be found in those inanimate objects which the imagination had infused with animistic properties, with make-believe life. Railway trains provided excellent opportunities. Train-spotters, just like fishermen, deer-stalkers and grouse-shooters, have first to calculate where they are most likely to find their quarry and must then contrive to be in the right place at the right time. It's the same sort of predictive competence which brings the same sort of reward to them all, at least in the first stage of the hunting process.

Perhaps you can now understand why I took the trouble to acquaint you with some details of my experiences in the bomb disposal squad (Chapter 4). The recovery and subsequent destruction of a large live bomb activates all the manifestations of the hunting instinct and satisfies a deep and universal craving in what, to most people, might seem a rather unusual way; but again, the satisfaction of the experience is greatly enhanced if one can persuade oneself to go along with a measure of animistic self-deception. The bomb becomes an adversary and one moreover which has a sort of capacity to fight back, not altogether unlike the capacity of a tiger or an elephant, which can easily kill you if you make a careless mistake. The squad is a collective band of communal hunters, dependent for their success not only on their individual skills but also on their collective, co-operative efforts. A more usual expression of the same thing is to be found in organised competitive sport, in which I personally never achieved more than a modest success, very modest in comparison with the prowess of my father.

Individuals vary in the relative strength of their desires to act individually and gregariously. We all value privacy as well as social involvement, and in my own case the former has generally tended to prove the stronger. I was able without difficulty to learn how to enjoy my own company, for example in my exploration of railways while my school friends pursued other more conventional occupations, or in those lone cycle rides at Oxford. There are many people who could simply not endure a whole winter's day pruning trees in an orchard without speaking to a soul. For me it was more than tolerable; I positively enjoyed it. I sometimes wonder whether I picked up from my father the aversion which he showed for large gatherings at camp sites and other holiday venues. On such occasions his attitude seemed to verge on the misanthropic, though generally his caring for, and involvement with, other people was, in fact, a conspicuous part of his nature.

Be that as it may, I find myself not powerfully attracted by some of those social activities which seem to be almost indispensable to some of my friends. I am not, for instance, a 'pub man'. The English pub lunch is fine. It's a purposeful activity usually offering simple food, acceptable to an unadventurous taste, at a reasonable price, and, especially if one is out walking, it provides a cosy element of refuge. Midday pubs also tend to be uncrowded; but the buzzing, bustling, smoke-polluted atmosphere which some people will endure with apparent pleasure for a whole evening only becomes tolerable for me if enlivened by a Dixieland jazz band. For that I would endure much!

The main differences between us all as observers probably arise, not so much from what we look at, or even from what we notice in what we look at, but from the meanings we attach to what we notice. This brings us back to the very heart of survival behaviour. If the acquisition of environmental information is of central importance, it's what we make of it that gives it that importance (Kaplan and Kaplan, 1982). First and foremost we need such information to enable us to find our way around. In a literal sense this means being able to understand where we are in relation to all those things around us which have a potential bearing on our behaviour. In a modern civilised society we may be able to manage quite well without a keen sense of location, just as we can survive comfortably with a sense of smell far inferior to that of many other creatures. We can negotiate quite complicated journeys if we can find our way to the bus stop and let the driver do the rest. Only when we have to make some critical decision which involves an awareness of our orientation do we need to be as good at it as other creatures need to be all the time.

One of the ways in which I differ from most people is in having an exaggerated awareness of direction (not always accurate but always 'felt' as a component of the landscape), as I have earlier had occasion to remind you (Chapter 5). Orientation in its most literal sense means 'a turning to the east'. We use the word less precisely to mean an ability to order our movements in relation to our surroundings, and I developed an above-average competence in this very early. Even in my dreams I'm nearly always conscious of facing a certain way. Perhaps the most cogent expression of this constant awareness of direction can be seen in my need to imagine what lies over the horizon even though, again, I may be very wide of the mark.

Placing things in their proper relationship to the points of the compass is simply one of many ways in which we classify the objects we observe. In other systems of classification, for example those of a mathematical or statistical kind, I have always been far less competent. Even in orientation my skill doesn't begin to compare with that of many species of bird which, using the sun, the earth's magnetic field, and other clues, can navigate for thousands of miles and arrive at the very nesting-site which had been occupied the previous year, though the belief that such a skill has been entirely lost by the human species may well have been much exaggerated (Baker, 1981).

Classification is an essential part of the process of making sense, which in turn is essential to survival, but it's only a beginning. In the perception of their environment all creatures need to be able to recognise not simply what something is (i.e. to what class it belongs), not simply what it means, but more specifically what's in it *for them*. This is called 'affordance', a term invented by Gibson (1979). Classification doesn't have to be a complex intellectual undertaking such as can only be achieved by the human brain or its substitute, the computer; all that's required is a recognition that a perceived individual or object belongs to a certain class whose general characteristics are understood. It's essential, however, to make that recognition quickly and correctly, so it shouldn't surprise us that classifying things, like so much survival behaviour, is a source of pleasure in its own right.

In my own environmental perception I seem to have shown a certain inconsistency in my ability to classify. Not just in assigning directions to the points of the compass but in a number of other ways I was fairly systematic, as is shown by my devising a highly original scheme for the arrangement of numbers (Figure 5) at a very early age; and, while I never managed to identify makes of cars as efficiently and reliably as my children could at the age of five, there was a time when I could pick out a railway engine half a mile away and attribute it to its proper class with unfailing accuracy. Certain other kinds of categorisation were also more important to me than they are to most other people, such as maintaining the distinction between sweet and savoury in food. On the other hand it's not to say that I'm content to accept all classifications as they're customarily presented, and my rebellion against the 'package deal' suggests that I attach importance to making my own classifications to suit my own needs and my own system of values. Nowhere is this more clearly demonstrated than in the process of making my own model of reality - my own world.

The ability to manipulate physical matter into a particular shape or structure has an obvious role in survival behaviour. Perhaps it can be seen in its most common form in nidification or nest making. Most birds, for example, have the capacity to assemble the right kind of raw materials, collected in the vicinity, and to build a structure which conforms to the general pattern of nest appropriate to the species. Some of these are of amazing complexity and involve a high degree of skill. The birds build them, not because of any rational understanding of the principles involved, but simply because, at a certain time of the year, they experience a desire to do so.

When we human beings make things we also are motivated by a comparable desire. We may indeed have some ulterior motive (and our superior thinking capacity may enable us to be more conscious of the reason), as when we fashion some implement to perform some particular task; but often we make things for no ulterior purpose we can consciously recognise. Some people go so far as to say that an object made for some ulterior purpose cannot be a work of art, since art itself cannot have purpose, but they would have to contend with no less an authority than Sir Joshua Reynolds. His are the words carved above the main entrance to the Victoria and Albert Museum in London: 'The excellence of every art must consist in the complete accomplishment of its purpose.'

That purpose may be simply to give to the creative artist the pleasure which comes from the act of creation, or to others the satisfaction of observing or in other ways interacting with what the artist has created. In all the arts imitative skills are necessary not only to produce images of existing objects but also to realise images conceived in the imagination. The painter of portraits, the maker of pots, the creator of abstract designs, the composer of music and the writer of verse are all practising skills like the nest-making birds because that's what they want to do. They're rewarded by the pleasure which comes from success and they're frustrated by the sense of failure which ensues when their skills are inadequate for the task they've attempted.

When I look back on the satisfaction I derived from pruning fruit trees (Chapter 4) it makes a good deal of sense to see it in this light. Starting with the shape of a wine-glass I was expected so to interfere with a growing plant as to make it likely that it would grow into that shape. The fact that it was unquestionably a utilitarian act didn't prevent it from being also an aesthetic experience. The pleasure came from practising and improving a skill which I wanted to be able to demonstrate if only to myself, and if one looks for a reason why, at least a partial answer is that it's in my nature to want to achieve efficiency in those creative tasks which I undertake, and that, if it were otherwise, my nature would be at variance with that of a creature motivated to survive. If you remember those examples of creative activity given at the beginning of Chapter 5 I think you'll find that the same explanatory approach is equally applicable to them all.

Another aspect of behaviour which has important implications for survival is a capacity to predict. Many catastrophic events are prefaced by warning signs, which, if we can perceive them in time *and* understand their significance, may enable us to avoid the consequences of such disasters. So we should expect the successful practice of prediction to be accompanied, like other survival behaviour, by sensations of pleasure, and it's interesting to note that animal experiments which test a creature's capacity to predict unfamiliar endings to familiar processes go under the general name of 'rhyme'.

It's difficult to escape the conclusion that the satisfaction which generations of poetry lovers have experienced from rhyming metrical verse is derived from the same source. The metre sets up a rigid framework of rhythm into which the word sequences are fitted, and, as the rhyming conclusion of a couplet is approached, it suddenly dawns on us how it's going to finish just before it does! What we're practising in an exercise of verbalisation is an extremely sophisticated development of what rats and mice are doing when they're being trained to anticipate circumstances which are similar to, but not identical with, other circumstances previously experienced. In this way it may be possible to throw bridges, however flimsy, across the gap which seems to separate so widely the fundamentals of survival behaviour from the most sophisticated products of creative art.

Of course the acceptance of strict limitations of metre and rhyme implies also limitations on our freedom to express the fruits of the imagination, and many people have found, particularly in the present century, that the price is too high. Whether they settle for metrical regularity, with or without rhyme, or whether they go for uninhibited freedom, depends very largely on what sort of people they are, which in turn depends at least in part on the sum of their experiences. I've suggested that, in my own case, these experiences have inclined me to follow the former course. Perhaps I found in Edward Fitzgerald and A. E. Housman the same sort of pleasure which those rats found at a more primitive level in the rhyming experiments of the zoologist's laboratory!

If this analysis of my taste in poetry is correct it would provide also an explanation of my taste in music where a fairly strict system of rules also applies (Chapter 6). A radical departure from the laws of classical harmony leaves me lost; but these laws themselves are not the product of arbitrary fashion. They're rooted and grounded in the laws of physics. 'Harmony' is produced when air is made to vibrate at two or more frequencies, each of which bears a very simple ratio to the other or others. The more complex the ratios the more dissonant the sound. Each individual seeks a compromise between two much complexity, which confuses, and too much simplicity, which bores. For my part I'm content to eat the same food day after day as long as it complies with a few simple rules, but I tend to be put off by more elaborate dishes which break those rules, so I'm not surprised to find that my taste in music can be described in very similar terms. For me a Welsh hymn tune, sung by a male voice choir from the mining valleys, can be as moving as a classical symphony and vastly preferable to a 'serious' work which has no recognisable rhythmic scheme and consistently breaks the rules of harmony; furthermore it can stand repetition, like chicken and chips!

Is it so surprising, then, that anyone who finds himself at home with these formal but simple frameworks for pleasurable experiences in poetry and music (and food) should also feel a need for a similar type of framework for environmental perception? The omnipresent sense of orientation into which, for me, every episode of perception has to be fitted, is just such a framework.

A central theme in environmental perception is what we may loosely call 'association'. It can take many forms and I've already given examples of several, for instance in connecting landscape with music (Chapter 6). We project on to some new experience some attribute or attributes of an old one, often quite sub-consciously. It may be a highly personal association, like that between a hymn tune and the Mouth of the Red Sea (p. 156), incomprehensible to anyone except me, or it may have some more widely understood basis, such as that which links the styles or fashions of some period, which anyone versed in the subject could easily pick up. The phallic chimney of a Wigan colliery, the middle-class elegance of the surroundings of Shrewsbury School (Figure 49) and the coniferous plantations of Sennowe Park (Figure 22) contained a common element which I could recognise as a linking *motif* long before I knew anything about late nineteenth-century architecture or landscape design. They induced a similar feeling, though in appearance they could hardly be more different.

It's this kind of association together with the idea of classification already discussed, which enables us to enrich our understanding of the 'meaning' of an experience. It helps us to give it a context. It tells us more than we can actually observe, though what it appears to tell us isn't necessarily true. Particularly when it's been sieved and sorted by the memory it may present us with a very different image from that which was projected on to the eye. It's to this that we owe our irrational prejudices, such as I experienced towards the Fens, as well as the phenomenon of nostalgia. The latter, as I'm sure you'll have noticed, rarely lies far below the surface in this book.

FIGURE 49. LATE VICTORIAN MIDDLE-CLASS RESIDENTIAL ELEGANCE.
View outside the Moss Gates, Kingsland, Shrewsbury. Photo by the author, 1986.

This view of environmental perception as being first and foremost a warning system to alert us to potential danger may take us a little way towards understanding the inclination we all have to allow a measure of animism to enter our image-making when it can have no strictly rational justification. Perhaps it springs from the notion that animate objects are more likely to be dangerous than inanimate ones, and that therefore, until we have had the opportunity to satisfy ourselves that a newly perceived object is inanimate, it's safer to assume the opposite and treat it as if it were potentially hostile. All those societies which we condescendingly describe as 'primitive' seem to work on the principle that, if there's any doubt, it's better to play for safety and assume that rocks and trees and rivers are less likely to take a dislike to people who behave towards them with due deference. Of course we're in a position to know better, but knowing better doesn't always stop us from eating more than is good for us, coating our lungs with nicotine or falling in love injudiciously, so why should it stop a small boy from saying good-night to the 'animal' on the Stibbard turnpike, even if he knows it's only a couple of trees, or a big one from getting emotionally involved with a heavy-breathing Stanier Pacific pulling a dozen loaded carriages up the 1 in 75 gradient over Shap Fells?

What's needed to bring to life an inanimate object like a mountain is the kind of imagination which can convincingly handle literal and metaphorical interpretations of the same picture. An imagination which has no difficulty in seeing a horizon as an internal boundary within a single terrestrial surface, part of which can be seen and part of which cannot, ought not to have too much difficulty in handling the concept of a mountain whose visible features come from the real world and whose personality comes from the world of fantasy. The vividness of the former is dependent on the immediacy of direct sensual perception, but this is not inconsistent with the idea that a fertile imagination can invest it with a different kind of vividness, a different kind of 'reality'.

Perhaps the most exciting moment in any exploratory act is that at which two images fuse into one, the moment when the world created in the imagination by the anticipation of what lies round the next corner suddenly becomes the world perceived by the eye. Invariably there are adjustments to be made, revisions of partially correct images, of half-truths which, being the best sources available, have had to serve us until the moment of revelation. We may not all express it in exactly this way, but I believe some such experience is familiar to us all and that we share a universal sense of enjoyment when it happens; and this matching of what is expected with what is perceived is, as we have seen, an important part of the machinery we employ to avoid disaster.

Every process of discovery takes place in stages at each of which we experience satisfaction. The awareness of something unusual is the beginning; then we may go on to recognise more precisely what it is, to place it within our classificatory system. From there we may go on to understand that it has some particular significance *for us*. Then we realise that we can project the situation forward and work out what this may mean for us this afternoon or tomorrow or when we reach some distant destination. At each stage something is brought into our understanding which, until that moment, was outside it.

This is certainly an area in which we need to recognise important differences between ourselves and other species, but modern zoological research is increasingly pointing to the fact that many of these differences are differences of degree rather than of kind and that hard-and-fast lines of demarcation have become extremely difficult to draw. It's no longer impermissible, even in scientific circles, to speak of an animal 'thinking' or even 'reasoning'. To say that it's incapable of predicting what will happen next Thursday is not to say that it has no powers of prediction at all.

One constantly-recurring problem is that, once we have made the reconciliation between what we expect and what we actually discover, the predicted and the revealed, and derived the satisfaction which comes from doing so, we can't 'unknow' the answer. We can never repeat that particular experience with the same expectation of reaping the reward of discovery. We therefore relish the fact that there are plenty more similar kinds of experience still waiting to be enjoyed, plenty of things still to be discovered, plenty of raw material for future exploratory successes. There is also something to be said for spinning out the phase of anticipation, building up the sense of expectation and delaying the moment of truth. Quite a lot of evidence has now been amassed to suggest that 'mystery' plays a special role in the enjoyment of landscape. Stephen and Rachel Kaplan have made a particular study of mystery and have shown that it frequently figures strongly in preferred landscapes (Kaplan and Kaplan, 1989, text and bibliography).

A problem which we often encounter when we try to rationalise our ideas about landscape is the difficulty of establishing logical connections between landscapes which are visually very different. It's easy enough to see why similar emotions should be evoked by the Alps, the Rockies and the Andes, but at the level at which we're now enquiring it may be necessary to look for associations between unlikely and apparently very dissimilar scenes. By way of illustration let me remind you of two examples which I've already mentioned. The first is the view of Helvellyn from Patterdale (p. 37) when rock and mist, mountain and cloud combined in a confusion which was itself delightful. The second is that type of visual experience (p. 167) which features small bodies of water fringed by green swards, almost level with the water surface, beneath which my imagination creates a world inhabited by populations of unseen fish. It may at first be difficult to see any connection between the two images, which could hardly be more different, but the element of 'mystery' is capable of inducing, at least in me, a kind of common feeling which links the two.

A particular kind of experience in which this element of mystery is always important is the dream (Chapter 5) because there's always a measure of confusion between those real scenes about which our senses have informed us and which furnish the raw material for making the dream, and those events and circumstances which originate in the dream itself. If the dream has been unpleasant the moment of waking is attended by a sense of relief, but nearly all my dreams are pleasant, or at worst neutral, and if I can remember anything about them when I wake up, as I often can, I look back on them with a kind of nostalgic desire, if only for the landscape images which have been lost. In all this there's a pervading element of 'mystery' and a feeling of longing fully as potent as any I have experienced for real places.

If, as I have suggested, environmental perception is a central part of the mechanism of survival, a mechanism which has evolved to ensure that, like other creatures, we are drawn towards those kinds of environment which are likely to prove more conducive to our survival, this raises the crucial question of what that optimum environment is. Put in another way, does the species *Homo sapiens* have a natural habitat in the sense in which the phrase can be applied to other species?

There are several levels of particularity at which we can aim the question. Let's look first at the most extreme argument which postulates that our own species does indeed have just such a natural habitat and that it is to be found in that area in which the anthropologists broadly agree the hominids first appeared, namely the African savannah. The idea that some such landscapes possess a kind of atavistic power of attraction which still persists in modern man causes as much hostility in some quarters as did the publication of *The Origin of Species* to the English bishops; nevertheless it is being very seriously argued, for instance, by Orians (1980 and 1986) and Balling and Falk (1982).

There isn't room here to develop this 'savannah theory', but let me refer to just one line of thought which touches directly on the aesthetics of landscape design. During the seventeenth century in Western Europe the landscape architects were busy fulfilling the need to express symbolically the supremacy of man over nature, as well as that of the landowner over his tenants and the 'lower orders' of society, through the use of formal geometrical figures, straight lines, rectangles, circles and so on. Everything was made to conform to order. By the eighteenth century, however, the supremacy of man over nature was beginning to be taken for granted, and, under the liberalising influence of The Enlightenment, more progressive ideas were already paving the way for the Romantic Revolution with its renewed emphasis on nature, the individual human spirit and the relationship between them, and querying the uncritical acceptance of a social structure supposedly based on divine authority. The new thinking challenged the landscape architects to develop new forms. So what was the pattern to which they turned? Still grass and trees, but now arranged in untrammelled freedom with a wide enough spacing of the trees to permit the voids to become as important as the masses. The eighteenth-century park may have been groomed, polite and very civilised, but it was nevertheless powerfully suggestive of an ancestry ultimately going back to the savannah, with great emphasis, wherever possible, on that one commodity not universally available in the savannah but essential to make it habitable for man and beast alike, namely water.

When I look back on my own preferred landscape types (Chapter 7) I find that many of the places which have most strongly attracted me can be seen to have at least some affinity with the savannah. That first impact with Holkham (p. 78) brought me face to face with as fine an example of the eighteenth-century English version as I could wish to see. But there were other variants, my favourite heathlands, for example, which, though differing greatly in tree species, in ground cover and in the pattern of spacing, resembled the prototype at least in a distribution of voids and masses which afforded a wide variation in depth of visibility, the view sometimes being closed quite near to the viewpoint and sometimes extending to the far distance.

The argument that the savannah is the natural habitat of our species and that we still react to it by some spontaneous machinery of perception which has been passed down through the evolutionary line as what Jung would have called an 'archetype' may seem too far-fetched to be convincing. It is particularly distasteful to those who are emotionally disinclined to believe that aesthetic experience can be explained by any hypothesis which science has to offer; and I will confess that I have never myself gone so far as to urge it, not having the necessary technical knowledge.

The validity of my own arguments does not depend on it, though clearly they could only be strengthened if it were confirmed. Rather I look for support from those more general principles which are described in this chapter and which are to do with such things as seeing, hiding, sheltering and escaping, and which, being basic to the adaptation of a creature to its surroundings, have to work in any environment whether of a kind familiar to our primitive ancestors or not. In fact most of the surroundings in which we have to put these principles into operation are of a very different kind, and if we've replaced the environment, to which our habits of perception are still instinctively attuned, by another, it means we're using a machinery which has evolved to enable us to live in one world for the purpose of trying to survive in another. Small wonder if we have problems!

None of us would want to live in the Stone Age, but we do need periodically to re-visit at least some symbolic representation of that world of nature which for our ancestors was an integral part of everyday life, and to recapture a sense of intimate interaction with it. That's why we bring plants into our city streets and even into our houses, why we bring water, where possible, into our urban parks and even our gardens, and why, if the opportunity presents itself at an acceptable price, we like every now and then to leave the landscape of civilisation and to surround ourselves with something more closely approaching the landscape of nature.

I use the word 'approaching' advisedly because it implies a *degree* of wilderness. It may help if we think of a continuum having as its two poles the wilderness at one end and its totally artificial replacement at the other, the natural jungle and the concrete jungle. We can conceive of the possibility that, somewhere along that continuum, there's an optimum point which represents the ideal and perfect landscape, but we must qualify this conception in the light of experience which tells us that it's more realistic to think of an optimum range within which our organs of perception are most likely to find satisfaction. That range will be different for each of us. Furthermore the preferred point within it will vary for any one of us from one time to another, depending on all sorts of variables, whether we're tired, whether we have something else on our minds, whether what we see reminds us of some previous experience, and so on. Having moved for a while along the continuum towards the wilderness end, we shall find that there will come a time when we're ready to retrace our steps towards the comforts and opportunities which we can only find in civilisation. After the *piste* the *aprés-ski!*

This was the story of Thoreau when he went to live by the side of Walden Pond in 1845 (Anderson, 1968). There he wrestled with that ambivalence which he felt towards the forest on the one hand and the civilisation which he enjoyed among his literary friends in Concord, Massachusetts, just down the road, on the other. You will know by now that my story has been very different from that of Thoreau, but the conflict itself has been essentially the same, as I believe it is for all of us. My Walden Pond has taken many forms. Landscapes wholly unaffected by human interference can no longer be found in England, but those mature, humanised landscapes, for which it is famous, afford many opportunities for making that contact with the world of nature at, so to speak, an intermediate level of intensity. For more powerful sensations of wilderness I have had to look further afield.

I have given you only the barest outline of what I called 'Habitat theory' and 'Prospect-refuge theory' when I first proposed them (Appleton, 1975), simplified catch-phrases intended to encapsulate a way of looking at landscape; and the point I want to emphasise once again is that, just as our habits of perception and our tastes in landscape are influenced by our life experiences, so also are our attempts to explain these things. None of us is a free, independent and objective thinking-machine; there is always the probability that what we have become will incline us towards some particular kind of explanation, some particular line of reasoning, and I think you will have no difficulty in seeing the connection between my own theoretical ideas and at least some of the autobiographical details which I have shared with you in the preceding pages. If not, let me suggest a few.

The first thing that must strike you is that these theories are deeply grounded in an evolutionary view of human behaviour, and my almost life-long admiration of Charles Darwin can hardly be irrelevant to this. Secondly, my early disillusionment with the teaching of the Stoics which had been invested with the status almost of an axiom by those concerned with my education, and the growth of a sneaking admiration for the hedonistic philosophy of Epicurus (p. 191), are entirely consistent with my eventually arriving at a view of aesthetics based on a central role for pleasure as a driving principle.

A third consideration is my concern with the idea of 'order'. In music and in poetry a fairly rigid system of order seemed to be characteristic of what I most enjoyed, and I attributed this, at least in part, to those classical values to which I was exposed throughout my school-days (p. 192); yet when I came to look at landscape I found that the informal parklands of the eighteenth century were often more enjoyable than the formal designs of the seventeenth. It seemed, therefore, that, as a criterion of aesthetic excellence, 'order' was neither here nor there. Where I had gone wrong was in looking for the wrong sort of order in the wrong sort of places. I had thought of 'order' as a property of harmony and rhythm in music, of rhyme and metre in poetry and of geometrical regularity in landscape. This is indeed the truth, but not the whole truth. The system of order which Darwin disclosed was to be found in the laws of the biological sciences, and included within these were the principles governing habits of environmental perception. Here, it seemed, were to be found more consistent pointers to a set of aesthetic principles, no less based on 'order', which would begin to make sense for me.

Fourthly, I earlier suggested that my precocity in orientation was encouraged by certain circumstances in my childhood, and I explained how I tended to prefer finding out about the world by direct exploration rather than by reading about it in books, and, fifthly, possibly arising out of this, I admitted to finding it easier to handle concrete than abstract ideas. If you look back over the present chapter I think you will see many ways in which my efforts to explain the aesthetic enjoyment of landscape reflect all these interests, tendencies and idiosyncrasies.

Finally, the fact that I have been driven to devote any thought at all to the explanation of these phenomena brings us back to the thesis that *Homo sapiens*, like all other creatures, is an exploring animal, and that, given his much greater powers of reasoning, it is in his nature to want to go beyond the exploration of his physical environment and to pursue the understanding, not only of what he perceives but, if he continues his quest long enough, of how and why he perceives it. And if we go on to ask what it is that drives him to do this it is that same pursuit of the pleasure which comes from success. As Virgil put it, *Felix qui potuit rerum cognoscere causas.*

10 POSTSCRIPT

'I travelled among unknown men
In lands beyond the sea.'

The process by which an individual takes the rudimentary system of instinctive attitudes and responses inherited from his or her parents, and shapes them into a pattern of habits and preferences is both gradual and lifelong; but it is punctuated by a series of steps, (what, after all, does 'gradual' mean?), some of which may mark significant changes in the way landscape is perceived, interpreted and enjoyed. In my own life the stage described in the previous chapter, when I began to formulate theoretical explanations of environmental aesthetics as I had come to understand them, was the most important single step in the process. It changed my thinking habits and brought me into contact with a large number of other writers working in the landscape field. It must therefore be expected to represent something of a watershed which, once crossed, opened up new perspectives.

In the light of this it may make a fitting conclusion if we look briefly at two questions. How did the changes in my own thinking, described in the previous chapter, affect my relationships with, and attitudes to (a) the new people, and (b) the new places I was to encounter?

It would be quite impossible to discuss either of these topics in full, because every person and every place involved was unique, furthermore, if I were to attempt to do so, this would soon become a catalogue of personal names and placenames, tedious in the extreme, not least for any readers not familiar with either; so I must resort to very broad generalisations, choosing particular examples to illustrate particular points, rather than to indicate my assessment of their relative importance. (This to placate my many friends whose omission does not imply any ranking of importance!)

In Britain most of the people working in the broad field of landscape aesthetics approached it through the arts rather than the sciences. This can be seen not least among my fellow-geographers, who comprise a substantial proportion of those who have gone into print. The discipline is (regrettably) divided almost constitutionally into physical and human geography, and, while a few physical geographers, such as David Linton and Edmund Penning-Rowsell, have made important contributions to the study of landscape evaluation and landscape taste, they are very much the exceptions.

A major centre of research into the aesthetics of landscape was to be found in the Department of Geography at University College, London, where it grew out of a strong line in historical geography, encouraged, but not initiated, by Sir Clifford Darby. Hugh Prince has for years been studying the great parks associated with English stately homes, and reading his work had been one of the stimuli which had originally caught my own interest. By the middle 'sixties he had produced two seminal papers in collaboration with David Lowenthal (Lowenthal and Prince, 1964, 1965), who shortly afterwards joined him at University College. Lowenthal, an American, had academic qualifications in history as well as geography, and this shows in most of his writing (Lowenthal, 1985). He had a wide knowledge of the academic scene in the United States and a keen understanding of the distinctiveness of national cultures. How many Americans, one may ask, on going to live in England, would embark on a process of acclimatisation by taking up church bell-ringing, as Lowenthal did at Harrow-on-the Hill?

Other lines of interest burgeoned in the Department, for example with Jacquelin Burgess and her collaborators, which I have referred to in the Preface (p.x). Here the 'cultural' input was contemporary, but in general the link between landscape aesthetics and historical geography remained strong. One of the most active products of the department, Stephen Daniels, has pursued a similar line in Nottingham, sometimes working with Denis Cosgrove, who went, from an Oxford background, to the neighbouring University of Loughborough. (See, for example, Daniels and Cosgrove, 1988.) My involvement with the Landscape Research Group made for me many personal friends in Britain (some of them being concurrently academic adversaries!), whose work, with few exceptions, has been heavily weighted towards the cultural side.

Individual geographers in Britain have concerned themselves with particular aspects of landscape aesthetics. For example, Douglas Pocock at Durham (landscape and literature) and Peter Howard at the University of Plymouth (landscape and painting, Howard, 1991). Outside academic geography there has been some activity from, for example, the psychologists, notably in the University of Surrey, but it has been very much a minority involvement. Consequently it is perhaps not surprising that many of my closest contacts have been with researchers in North America, where the situation is different.

America has not been short of scholars who have followed a 'cultural' line. One of the most prolific has been Yi-Fu Tuan, who, coming from an oriental background, collected a geography degree at Oxford before embarking on an academic career in Minnesota. But it would be much easier to find in the United States than in Britain advocates who have not only acquiesced in, but set out to press the case for the survival of a fair measure of the innate, as opposed to the learned, in our tastes and preferences for particular landscape experiences. Among the most influential have been Stephen and Rachel Kaplan, a husband-and-wife team of environmental psychologists in the University of Michigan. What particularly excited me was the discovery that, although we came from very different academic backgrounds and employed a different methodology, we arrived at remarkably similar conclusions.

For me a very particular significance attaches to the University of Washington and the City of Seattle, where there is an unusual concentration of individuals, owing allegiance to different disciplines and different departments, working together in the study of ideas related to what I have called 'habitat theory'. In an extreme form this has given rise to a course on 'Evolutionary Aesthetics' run by three faculty with very different academic backgrounds. Gordon Orians, an ornithologist holding a Chair of Zoology, has written the definitive work on the American blackbird (Orians, 1985). An interest in site-selection led him to draw comparisons between the methods by which animals, particularly birds, choose nesting-sites and the preferences which people show for particular kinds of environment (Orians, 1980, 1986). The second contributor to the course was Grant Hildebrand, a Professor of Architecture and Architectural History, (and one could scarcely have a more different academic background than that!). He took the idea of habitat theory and applied it to the domestic architecture of Frank Lloyd Wright (Hildebrand, 1991). (Another recent example of the application of prospect and refuge to architecture, incidentally, can be found in Brian Hudson's work in Brisbane on the verandah. Hudson, B., 1993.) The third member was Judi Heerwagen, Professor of Environmental Psychology. This is precisely the sort of inter-disciplinary collaboration which seems to me to be essential if we are to struggle out of the straight-jacket imposed by a system of rigidly separated subject areas, and I found it very exciting.

My initial contact in Seattle had been with the department of Landscape Architecture, where a number of the faculty had shown interest in the habitat theory approach, among them Richard Haag. Richard, having won international renown for his imaginative treatment of the site of the disused gasworks in Seattle, went on to design part of the Bloedel Reserve on Bainbridge Island in Puget Sound, a project which he described as 'the ultimate distillation of prospect/refuge theory' (Frey, 1986, 56). Needless to say I was gratified when, in 1986, as the gasworks site had done previously, it won a President's Award of the American Society of Landscape Architects.

The culmination of my association with Seattle came early in 1988 when I went to the University of Washington to deliver the Jessie and John Danz Lectures. By this time the 'nature versus nurture' argument had become so polarised that, if one ventured to argue on either side one was in danger of being assumed to deny the validity of the other! So I decided to use the lectures as a forum for arguing the absurdity of this dichotomy through symbolism. I distinguished between 'cultural' symbolism, where symbols, like St Peter's keys, a bishop's crook and mitre, or the lily of purity, have been chosen by human agency to represent individuals, offices or abstract qualities, and 'natural' symbolism, where the representation of ideas by objects occurs quite naturally. For example, precipices naturally suggest danger, good viewpoints naturally suggest strategic advantage, and so on. The symbolic association does not have to be artificially set up., To a true Darwinian, of course 'culture' is simply the most recent phase in a single evolutionary process lasting many millions of years. We should not think of it as providing an explanation alternative to nature; it is a part of the natural order. The fact that it is of immensely short duration in comparison with the whole does not prevent it from being by far the most important phase, but it is still a part of that whole.

The lectures were subsequently published (Appleton 1990), and in the following year the reconciliation between the 'heredity' and 'environment' schools was carried a stage further by Steven Bourassa (1991), who proposed that habits of perception and landscape taste are the products of 'biological laws', 'cultural rules' and 'personal strategies', never acting alone but always in some sort of combination. This seems to me to be a convenient formula to solve a problem which should never have been allowed to arise.

On my first visit to Seattle in 1978 I was giving an informal talk to a small group of faculty and graduate students in Landscape Architecture and noticed among them a much older gentleman, tall and distinguished, who was afterwards introduced to me as Prentice Bloedel. He told me he had just read *The Experience of Landscape* and invited me to his residence on Bainbridge Island where we could have some further discussion on landscape. Two or three years elapsed before an opportunity arose, but from 1981 onwards I stayed with Prentice and Virginia on several occasions, both on the island

and in Seattle and we became close friends. It was Prentice who had commissioned Haag to do the work on Bainbridge Island, but he subsequently handed over the management of the Reserve to others, and Haag's design has been considerably altered since.

Prentice Bloedel would serve well to illustrate the proposition, in a different context, that our attitudes to landscape are largely shaped by our life stories. After taking a Liberal Arts Degree he had joined the business of Macmillan Bloedel and brought to this huge firm of timber merchants a policy, which was well ahead of its time, of planting at least one tree for every one cut down. He had lived for many years in British Columbia, and had acquired something little short of a passion for landscape, which we spent many hours discussing.

I afterwards learnt that Prentice had been introduced to my work by Charles Lewis, who again had arrived at a remarkably similar view to mine by way of a very different route. Charles was an arboriculturist at the Morton Arboretum in the western environs of Chicago and had written on the therapeutic properties of plants and of gardening, a theme which had an obvious relevance to my own views.

My involvement with the Landscape Movement brought me into touch with many other writers and researchers in America. If I had to pick out one name from a long list it would have to be that of J. B. Jackson, the founder, and for many years editor, of the periodical *Landscape*, not only because I was greatly influenced by that publication, and not least his own contributions to it, but also because it was he who, over breakfast at a conference in Essex, suggested that I should write this book.

Most of my visits to American campuses have been hosted by the landscape architects, among them the Universities of Oregon, California at Berkeley, Illinois at Urbana, and Georgia at Athens. North of the border I had a particularly fruitful stay in the University of Guelph, Ontario, where, for two weeks, I was given the title of 'distinguished Visitor in Landscape Architecture', a pretty flattering title for someone with no qualifications in the subject whatever!

An early inter-disciplinary group was to be found in Tucson at the University of Arizona, and this conveniently leads on to my second question, namely 'Had my own habits of perception been so changed that encounters with unfamiliar landscapes, perceived for the first time, induced different responses, intellectual or emotional?' While I was on a visit there my host, Ervin Zube, one of the leading figures in landscape research in America, introduced me to the Sonoran Desert. All the symbolism of prospect, refuge and hazard seemed to fit, but in quite new ways. This landscape has become universally familiar through Western movies, but how much greater its impact when one experiences it at first hand! Immensely long prospects combine with sparse and weak symbols of refuge grudgingly afforded by extraordinarily shaped *saguaros* and other lesser cacti.

It was through another American contact that I first saw the American 'South'. Charles Anderson was a retired Professor of English at Johns Hopkins University in Baltimore. Not only had he increased my understanding of the importance of Thoreau in the history of American attitudes to landscape; he was also the editor of the collective works of his kinsman, Sidney Lanier, 'the poet of the marshes', whose work I had used in exemplifying Prospect-Refuge Theory in *The Experience of Landscape* (Appleton, 1975, pp. 146-8). Iris and I had become good friends of Charles and Mary in England, and when we received an invitation to stay with them in Charleston, this created an opportunity to see something of the landscapes of South Carolina and Georgia.

Seeing a place for the first time had always been one of my most enjoyable experiences, but until the age of about fifty, although I had often wondered what was happening when I encountered a new landscape, I hadn't been able to suggest even to myself any coherent reasons which would explain my own emotional responses. My first visit to Australia, for example, at the age of forty-one (Chapter 7), changed my perspective of the world in a way that hadn't happened up to that time, but if the *feelings* it evoked were responding to some set of principles, I didn't know what they were. I've often re-lived those events (and in many cases revisited the places), and been able retrospectively to fit them into the kind of explanatory structure which I described in the previous chapter, but I've always been left wondering whether the initial impact would have been different had my instrument of interpretation, my brain, already contained such a structure, and, if so, how my first impressions of places might have been different.

All this is bound to be a matter of speculation, but when I next travelled beyond the confines of Europe I had already begun the first draft of *The Experience of Landscape* (though it was not to be published for another three years) so it was not unreasonable to suppose that I should be consciously applying those theories, which were at that very moment taking shape in my mind, to the landscapes I was seeing for the first time. Starting in 1972 I began travelling again to more distant places, extending my knowledge of the regional geography of the world.

In the summer of 1972 I made the first of these later journeys when I attended the Twenty-second Congress of the International Geographical Union in Montreal and first set eyes on the North American Continent. It was during this visit that I first saw the

Canadian Houses of Parliament, satisfied myself that they really were on the southern and not the northern bank of the Ottawa River (p. 121), and discovered once again how difficult it is to eradicate geographical misconceptions once they have become engrained in the mental map. A pre-Congress meeting in Toronto provided an opportunity to see a little of the so-called Canadian Shield, with its vast areas of lakes and forests, as well as Niagara Falls on the other side of Lake Ontario, and before going on to Montreal I spent a few days with Russell and Ruth Peck, American friends who had bought a derelict farm in the forested area some eighty miles west of Ottawa.

I think it was in the Shield Country that I began to be fully conscious of the challenge of testing my evolving theories against the reality of experience. I was particularly fascinated by a group of so-called 'cottages', weekend hideouts, fringing Charleston Lake in Eastern Ontario. Although almost at the water's edge they lay behind a thin screen of trees which partially concealed them from the lake. They were perhaps the first clearly identified examples of what I soon realised was one of the most ubiquitous symbols of refuge, the cottage in the wood, powerfully expressing the concept of 'seeing without being seen'. This led me, as soon as I returned to England, to search for similar examples in the history of landscape painting, and I soon discovered examples in almost every period. The seventeenth-century Dutchman, Hobbema, for instance, had been almost obsessed by precisely this association of cottage and forest. One example, chosen from literally dozens, appeared in *The Experience of Landscape* together with a photograph of the cottages at Charleston Lake (Appleton, 1975, Figs. 45 and 46).

My first visit to New York was a very brief one, and like all tourists I was fascinated by the largest symbol I had ever seen of the mastery of the human species over the material world. I remember thinking also that the streets bore a closer resemblance to natural canyons than to other streets as I had known them in English cities, and I learnt just how powerful can be the concentration of attention on a single distant vanishing-point when visibility is so rigidly bound by opaque lateral screens. But perhaps the message which came over to me most vividly was the impossibility of an individual ever being able to exercise in such a vast conurbation the kind of personal relationship with an open-air habitat such as I had known in childhood and which, for me, is implied in the term 'territoriality'. My mind went back to Stibbard and the rights of access which I exercised *de jure* over the garden and *de facto* over the churchyard and the surrounding fields.

I remembered one of those little gramophone records which used to live on the bottom shelf of the dresser in the dining room at Stibbard, and which told the story of the country mouse and the town mouse and I knew that circumstances must have made me different from those children who had been raised in the Big Apple. All the same, some years later I asked Kenneth Helphand, Professor of Landscape Architecture at Eugene, Oregon, who *was* raised in the Big Apple, to comment on a draft of this book, and we discovered that, different as our experiences had been, down at that fundamental level of response to the basic environmental realities, like being able to see, to hide, to escape, we could both find in our vastly different environments much common ground to which we could sympathetically relate.

Greatly as I enjoyed the excitement of being in New York City, it was with a sense of relief that I boarded the plane at Kennedy Airport, not, alas, with my camera. That had gone off with somebody else!

Later that year, in November, 1972, I paid a short visit to Hungary under an exchange scheme set up by the British and Hungarian Governments. My obligation was to give a few lectures in Budapest and Szeged and to talk with students. In exchange for these modest duties I received the most generous hospitality, too much, in fact. I was expected to drink as many tots of brandy as my hosts placed in front of me. This they began to do at about eight o'clock in the morning and stopped doing so only a little while before I went to bed, there being by that time nothing else I was fit for. When I recovered I was taken to see the sights of Budapest and I made excursions further afield into the Hungarian countryside. It was then that the Schubert Octet became an integral part of the Hungarian landscape (p. 156).

During that visit there was one incident which shows how I was still trying to fit my own landscape experiences into some sort of theoretical system. I was being driven through the Great Alföld, the Great Hungarian Plain, in the company of an English girl, resident in Szeged, who remarked that she found this landscape fascinating but uncomfortable, because, she said, there was 'nothing to hide behind'. In America earlier in the year I had heard David Lanegran, a geographer from Macalester College in Minnesota, read a paper on a novel by O. E. Rölvaag called *Giants in the Earth*, and this had set me thinking about 'the landscape of exposure'. I duly read the book which was about some Norwegian settlers in the plains of South Dakota and it described how the wife had neurotically failed to come to terms with the immensity of the Great Plains. One sentence I had remembered in particular. 'If life is to thrive and endure', she says, 'it must at least have something to hide behind'. I'm pretty sure it was then that I adopted the word 'refuge' to symbolise whatever it was that was so deficient in South Dakota and, to a lesser extent, in the Great Alföld.

It was three years later when I made my first proper visit to the African Continent. I had just set foot on its north-eastern extremity when I visited the Great Pyramids on my way to Australia in 1961, but in November, 1975, I was asked to act as external examiner in the University of Rhodesia, as it was then called. Direct flights to Salisbury (Harare) being suspended for political reasons, I had to travel via Johannesburg, and I therefore took the opportunity to see some of the landscapes of South Africa before flying north.

I particularly remember how appropriately the symbolism of prospect and refuge seemed to work in the Valley of a Thousand Hills, which one passes on the train as one approaches Durban from the north-west. Innumerable little tributaries, falling steeply into the main valley from either side, have cut their own tiny ravines which have dissected the lip of the valley into innumerable bluffs, hence the name 'a Thousand Hills'. The result was an alternation of seeing-places separated by hiding-places, vantage points succeeded by secluded recesses. This pattern, which stretched for mile after mile displayed quite dramatically the difference between prospect-rich and refuge-rich situations and the pleasing effect of their repetition in close juxtaposition.

Another memorable experience in South Africa was spending a night with some companions from the University of Natal in a hut in the Drakensberg Mountains. The stark horizon of Cathedral Peak to the west contrasted dramatically with the vast open views over the veldt to the east. I wrote a little poem, later included in *The Poetry of Habitat* (Appleton, 1978), in which I tried to contrast the concepts of prospect and refuge as expressed not so much in landscape features as in day and night. The huge panoramic view over the veldt, romantically coloured by the soft evening light, was gradually extinguished as little refuge areas of darkness slowly spread and merged into the single hiding-place of the night which eventually covered the whole landscape.

My task in Salisbury allowed me time to see something of the environs of the city and to make two longer visits. One took me to the Victoria Falls (Figure 46), where I enjoyed my most dramatic experience of the Sublime. When I had first seen Niagara three years earlier I was still in the process of working out the problem of reconciling danger with enjoyment. By this time, however, my thoughts on the symbolism of 'hazard' had come together, and even as I was walking slowly through the forest up to the safety limit and making immediate contact through all the senses (Chapter 7) with what lay beyond, I was now able to see myself as a creature in a potentially hostile environment testing where the boundary between safety and danger actually lay and deriving from the experience the most intense pleasure.

The other of these longer trips took me south to the ruin-covered hillock of Zimbabwe which was later to give its name to the whole country. After spending a night in one of the huts provided for tourists I visited the game part at Lake Kyle, a water-body, incidentally, which must pose serious philosophical problems for those who believe that man-made reservoirs must be intrinsically unattractive. There I saw many species of grazing and browsing animals which had come down to the waters of the lake in the early morning and were now beginning to retreat as the day grew hotter into the more heavily timbered parts of the savannah. This was a memorable demonstration of the delicacy of the relationship which must exist between herds of animals and their habitats if they are to exploit them to their own advantage and of the dominant role which water plays in the ecology of the savannah, and it was this image which I called to mind when Gordon Orians introduced me some years later to the 'savannah theory' of landscape aesthetics (p. 221).

At the beginning of 1980 I had the opportunity to spend six months as Visiting Lecturer at the University of Canterbury in Christchurch, New Zealand. My earlier experience of Australia had led me to want to see this country which was alleged to be in some ways so similar and in others so different. As a geographer I went with many questions of a more orthodox kind concerning its economic geography, for instance, and not least its railway system. But it was the prospect of seeing its scenery that most excited me, and again I found myself applying theories of environmental perception and landscape aesthetics to new visual experiences.

I'll give you one example. If you were to take a cross-section through the southern half of the South Island a little further south than Christchurch you would find the Southern Alps rising to summits of ten thousand feet and more, only a few miles back from the west coast. The high rainfall on the western slopes provides the conditions for the growth of thick forests while the vegetation in the rainshadow on the east is grassland. What is so exceptional is the suddenness of the transition from the one to the other. I crossed this boundary several times and was always struck by the instant change of feeling, of attitude, of emotional awareness, which I interpreted in terms of extreme manifestations of refuge and prospect, corresponding with the forest and grasslands respectively. Whether I would have reacted in the same way had I first encountered this landscape before I had formulated these concepts is a matter for speculation.

There is just one other episode which I should like to recall from this period in New Zealand. Over forty years earlier I had read *Erewhon* by Samuel Butler as a set book in English at Shrewsbury, Butler's old school, and I recollected how the earlier chapters describe a journey up a valley and over a pass into the mythical land of topsy-turveydom. This journey, I understood, was based on one which Butler had actually made with a Maori guide during the few years which he spent sheep-farming at a place called Mesopotamia beside the Rangitata River. On one occasion we visited the remains of Butler's homestead and the following weekend Mark and I took the car to a place now called Erewhon at the end of the road on the opposite side of the river. There we left it, climbed through the fence, dropped rapidly to the river, and, following its tributary, the Lawrence, with a copy of *Erewhon* in our hands, set out for the Butler Saddle, the pass at the valley head.

In his account of the journey Butler takes a number of liberties with his topographical description, nevertheless the general character of the valley with its broad spread of stones laced with swiftly flowing streams is depicted accurately enough as is the saddle which still bears his name. The gorge which the travellers encountered towards nightfall and where they made their camp, is rather like the waterfall in *Lorna Doone* in that it owes more to the imagination than to the original landscape, but there is a point where the valley is greatly restricted by a scrub-covered moraine which invades the valley from the eastern side, and, taking this to be the prototype of the feature described by Butler, we emulated him in choosing this spot for our first night's resting place.

In the whole valley there were no fences, no signs of domesticated animals and certainly no suggestion of a path. The sides of the valley were so thickly covered with thorn that the only practicable course lay along the stony river-bed which filled the valley bottom from side to side, some three quarters of a mile across, and every now and then we had to wade through the river itself (Figure 50), a task which Butler pronounced impossible except for men on horseback. With the river in spate, this could have been an accurate description. During the whole weekend, from the moment we left the car to the moment of our return three days later, we saw only two signs of humanity, a couple of vapour trails in the blue sky.

As the sun set and darkness enveloped the valley the little pool of light in the tent and the glowing embers of our supper-fire symbolised a diminutive outpost of civilisation in an area of total wilderness. We were equally excited by the sense of contact with the natural world on a vast scale and by that little beacon of warmth which told us we were members of that master race which had so far subdued it that we could penetrate its deeper recesses with every expectation of being able to find our way back in safety, if not to Concord, Massachusetts, as Thoreau did, at least to Christchurch.

On our return journey from New Zealand we first visited a number of good friends and old haunts in Australia and then made

FIGURE 50. THE APPROACH TO EREWHON. The Valley of the Lawrence River, South Island, New Zealand, with Mark Appleton carrying the tent (and much else!). The little shrub-covered moraine, where Samuel Butler and his Maori guide probably camped for the night, can be seen invading the stream-bed from the right in the middle distance. At the head of the valley the pass, now known as 'The Butler Saddle', from which the hero of *Erewhon* supposedly saw the mythical country 'over the range' (to quote the subtitle of the book), actually overlooks the headwaters of the Rakaia River. View to the north-north-east. Photo by the author, 1980.

FIGURE 51. HOMERIC LANDSCAPE, 1985. The harbour of 'Sandy Pylos', better known today as Navarino, site of the famous battle of 1827, seen here from the ruins of Nestor's Palace. Photo by the author.

briefer stops at several places in South-east Asia, Singapore, Kuala Lumpur, Penang, Dacca (as it was then spelt), Calcutta, Kathmandu and Delhi. Apart from a couple of shore visits to Aden and Colombo in 1961 this was my first contact with the cultural landscapes of Asia, and, brief and superficial as it was, it provided a great deal of food for thought.

In *The Experience of Landscape* I had argued that cultural differences, both in landscapes and in attitudes to landscapes, which may seem to be paramount, are often merely idiomatic expressions of the same underlying principles; and here in Asia, among cultures of great antiquity, I had ample opportunity to test this out. Everywhere the eye was bombarded with novelty, and one was tempted to wonder whether the built environment, in particular, reflected the application of principles which had anything in common with those of the western world. However, as the novelty wore off, it became clear that the really basic concepts, like indoors and outdoors, high and low vantage-points, nesting-sites and foraging-grounds, prospect and refuge, were equally valid, even though they found expression through wholly unfamiliar forms.

Meanwhile I had for some years neglected the Continent of Europe, but in 1985 I resumed contact with a visit to Greece. Our youngest son, Mark, and Tina Harrison graduated from the University of Hull in 1984 and decided to take up careers in teaching English as a foreign language. Apart from a few short periods in England, this meant working abroad, and it was through them that I found myself in parts of the world I should probably never otherwise have seen. Their first appointment was at a school in Katerini in Macedonia. With the exception of a few hours in Athens, while the *Orontes* was in port at Piraeus in 1962, this was my first visit to Greece. Iris and I joined Mark and Tina for their Easter holiday. Starting from Katerini we travelled through Central Greece, the Peloponnese and the islands of Mikonos, Paros and Santorini (Figure 36).

There was a time when my imagination had been much exercised in creating a visual Hellenic world as a kind of backcloth to those stories of antiquity which I'd been required to study (in Greek) for examinations; but that had been nearly fifty years before, so there had been a huge time-gap between the creative act of building this world, with all its fallacies and misconceptions, and the corrective task of matching it with reality. The stories themselves, of gods and mortals, I had largely forgotten, but not beyond the point at which they could still be revived; and it was a really exciting experience to stand in Nestor's Palace (Figure 51), looking south-west over the harbour of sandy Pylos, Homer in hand, and re-reading after all those years how Telemachus had arrived with Pallas Athene (in heavy disguise) in search of his father Odysseus at that very spot, and how Peisistratus, son of Nestor, had greeted them with an invitation to join the family barbecue on the beach. (Any doubts as to whether this was the precise site where these events had taken place I was happy to sweep aside!) If only I had known what the place *looked* like when I had waded through the Odyssey as a teenager, if only the tale had been told in the context of such a real landscape, how much more might the whole epic story have come to life!

Towards the end of 1986 Iris and I embarked on our most ambitious journey. By this time Mark and Tina were teaching at a language school in Buenos Aires and had invited us to join them again, this time for their long summer holiday. Having arranged to go as far as that I was able to combine it with a lecture tour incorporating universities, etc., in Argentina, Chile, Colombia and Venezuela.

Once more the new experiences, first contacts with new places, tended to be fitted into the aesthetic system which I had worked out and were therefore perceived with different eyes. So, for example, when I first saw the little chalet or *cabaña* where Mark and Tina had arranged for us to spend Christmas (Figure 52), I immediately recognised it as a powerful expression of a pleasing balance between prospect and refuge, not oppressively hemmed in by overpowering opaque vegetation such as might arouse a sense of hazard where it wasn't wanted, nor yet subject to an unacceptable feeling of exposure which is sometimes the price that has to be paid for a situation high in prospect values. The refuge symbolism of the *cabaña* was strengthened by that of the sporadic shrubs and trees in a highly satisfying combination.

This isn't the place to write a full account of the two months we spent in South America. Let me touch on just four topics or episodes which have a direct bearing on the theme of this chapter, which is that the way we perceive new landscapes can't be divorced from what our perceiving mechanisms have become through the sum of our previous experiences, and that, at this stage of my life, an important component of this mechanism was a tentative system of explanation of the aesthetic impact of what I perceived.

The first topic, then, is how difficult it is for rational thought to over-ride irrational but well-established images. My very early impressions of South America as a place where one could stand on the Andes and simultaneously see the Pacific and Atlantic Oceans to west and east respectively, (p. 118), albeit in the far distance, must have been rationally corrected by, at the latest, the age of six. Now its absurdity was to be fully exposed as we criss-crossed the Patagonian Desert on our way south. Even at Rio Gallegos near the east coast,

FIGURE 52. FIGURES IN AN IDYLLIC LANDSCAPE. Tina, Mark and Iris at Colonia Suiza, Bariloche, Argentina. Photo by the author, Christmas Day, 1986.

FIGURE 53. THE BEAGLE CHANNEL TO THE WEST.
Note the straightness of the channel, commented on by Darwin in *The Voyage of the Beagle*. Photo by the author, 1986.

by which time, on an atlas scale, the continent had tapered to almost nothing, and in crystal-clear weather, the mountains were still well out of sight beyond the western horizon over a hundred miles away.

So, had that childish figure, standing Cortez-like on an Andean peak and synoptically enjoying a view of the world's two greatest oceans been finally laid to rest by the irresistible powers of reason? I have to tell you, not only that it had not, but that I couldn't altogether regret it. An erroneous product of the imagination, exposed by rational investigation, may be rejected as a misleading falsehood and still cherished as a creative work of art!

My second topic takes us back to my schoolboy hero whose thinking formed the very foundation of my 'habitat theory' of landscape aesthetics. It was during the voyage of *The Beagle* (1831-36) that Darwin made most of the field observations which turned out to be crucial in the development of his theory of evolution. If the most dramatic of these had to wait until he reached the Galapagos Islands they proved to be the tinder which kindled a fire that had already been laid on the South American mainland, though it's true that over twenty years were to elapse before it burst into flame in *The Origin of Species*. The Cordillera of the Andes which, through most of Chile, is aligned from north to south, sweeps round at its southern extremity and assumes an east-west direction. A partial drowning breaks it up into a large number of islands of which Tierra del Fuego, the Land of Fire, is by far the largest, though by no means the most southerly. The Straits of Magellan, which separate it from the mainland of Patagonia, lie mainly to the north of the fold-mounts and describe an irregular, tortuous shape. By contrast the Beagle Channel on its southern side lies within, and is rigidly confined by, the folds of the mountains which impart to it a conspicuously straight alignment.

While we were staying in Ushuaia, allegedly the most southerly town in the world, we took a boat trip into the middle of the channel, and what chiefly struck me, apart from the magnificent colonies of birds and sealions, were the superb vistas to east and west along these stretches of water almost as straight as Fifth Avenue. When I eventually got home to England I re-read Darwin's own account of the voyage of *The Beagle*, and was particularly struck by his story of how he was invited to accompany the ship's captain, Robert Fitzroy, in a small boat to survey a part of the channel. 'With this beautiful weather', he wrote, 'the view in the middle of the Beagle Channel was very remarkable. Looking towards either hand, no object intercepted the vanishing-point of this long canal between the mountains' (Figure 53). I found a curious excitement in the discovery that Darwin, having been in the same place a century and half earlier, had been struck by precisely the same feature which so impressed me, and that it was to do with one of my favourite environmental experiences, being able to see a long way, a dramatic expression of the idea of 'prospect'.

My third topic, culled from a feast of visual experiences on this South American tour, relates again to the Andes but in a different role. It takes us back to Chapter 2 and to the feelings inspired by Rivington Pike as seen from Bradley Hall. You may remember how, as an outlier of the Pennines, it provoked a sense of seclusion, almost of protection, from the east. The idea of what I would now call 'refuge', which it symbolically afforded, was coupled with an invitation to the imagination to leap over it and explore the territory on the other side without forfeiting the protection it afforded, thus anticipating by nearly forty years the episode of *Pembroke-im-Pitztal* and by nearly fifty the rationalised concept of refuge and prospect.

I had since experienced this feeling many times, in Norway, for example, on the west coast of New Zealand and on the Pacific Coast of North America from California to British Colombia. In every case, the mountain barrier lay to the east. I now felt it very powerfully as I made my way north through a series of staging-posts up the west coast of South America, Puerto Montt, Santiago, Antofagasta, Arica, Lima and even Bogotá, which, although it stands on a high plateau over two hundred miles from the coast, nevertheless nestles right under a mountain ridge on its eastern side.

Further to the north-east Caracas is dramatically overhung by a wall of high mountains, but this time on its northern side, the trend of the Andes having once again swung round to an east-west direction so that the valley in which the city lies is open at its eastern end. I was keenly aware that, for some reason, the feeling of protection which the site afforded, powerful as it might be, was somehow different; it was not the 'Rivington Pike feeling'. So this set me thinking. Why should these barriers of hills and mountains have that particular emotional effect only when they lay on the eastern side? Why hadn't I felt it when I spent that night in the Drakensberg Mountains with the spectacular skyline of the continental watershed rising only a few miles away, but to the west? Why didn't it work in the Swiss and Austrian Alps? They were more than high enough, but the valleys generally ran east-west so the high barriers lay to north or south (or both), but not to the east.

I came up with two mutually-supporting hypotheses to throw light on what was by now a sixty-year-old phenomenon at least, and I was amazed to think that they hadn't occurred to me before. The first was that, because of the eastward rotation of the earth, all celestial bodies, the sun, the moon, the stars and the planets, whether perceived in the northern or southern hemisphere, appear to rise above the eastern horizon and to disappear beneath the western. (Remember the three moons dream?) Any animistic overtones which might be attached to these celestial bodies, therefore, (and a glance at the literature of the ancients will assure you that I was not alone in making such attachments!), whether malignant or benign, would direct one's awareness of a possible threat towards the east. The sensation of having the protective barrier to the west could never lead to a similar sense of protection against an on-coming hazard, because, if there were any celestial bodies visible above the western horizon, they would already be in retreat towards it and would soon disappear out of sight.

This phenomenon can be perceived anywhere outside the polar regions and lies within the experience of any sighted person, though I very much doubt whether most people would rationalise it in quite the same way. My second hypothesis, however, is more specific to my own individual history and relates to the time and place in which I first became aware of the points of the compass. When I first went to Stibbard as a toddler less than three years had elapsed since the signing of the armistice which had concluded the First World War. Those sepia photographs of the trenches were very much a part of my life during my first impressionable years. Had I been born a hundred years earlier, shortly after the Battle of Waterloo, I suppose it would have been the French who would have symbolised the enemy, and the potential source of danger might therefore have been associated with the south. But in the early nineteen twenties I was taught (not by my parents, who were deeply into Christian reconciliation, but largely by my peers), that the French had recently proved themselves to be the goodies and that if I was looking for the baddies it was to Germany that I must turn.

Germany, of course, lay to the east. In Norfolk the sun, the moon, the stars and the planets rose above a low, open, unprotected horizon like a progression of Teutonic invaders coming to spy on me out of the sky. They came, as it were, from the country where the people lived who had killed all those men from the village whose names I passed on the war memorial every morning when I went to fetch the milk. How comforting it would have been to have a range of hills covering one's eastern flank, even if, like Rivington Pike and the Pennines, they were no more than two thousand feet in height! The Andes, of course, are, at their maximum, about ten times as high and four thousand miles long, and that really is something to hide behind!

My last topic takes us to one of the remotest inhabited places on earth, arguably the remotest, if one excepts the scientific base on the South Pole itself. After I had given some lectures at a short conference in Santiago we headed west and spent six days on Rapa Nui, Easter Island, which lies in the South Pacific at roughly 27° South and 109° West. There you can imagine me standing on the rim of the crater of an extinct volcano, surveying the island from a fine vantage-point and trying to put my thoughts into some sort of perspective.

The crater of Rano Kau is rather more than half a mile across. The bottom is filled with water, blotchy with many patches of weed. To the south the outer edge of the volcano drops steeply a thousand feet into the blue Pacific out of which rise the tiny islands of Motu kao-kao, Motu-iti and Motu-nui (Figure 54). Iris, Tina, Mark and I have been joined by a guide and she is telling us in Spanish the story of the birdmen of Orongo, the primitive hole-in-the-ground village whose admirably restored remains surround us; how every year when the sooty terns came to nest on the islands there was a competition to see who could swim to the islands through shark-infested waters and retrieve the first egg, thereby earning for the head of his tribe the title of birdman for the year and glory for life.

As I listen to this story my immediate response is to wonder how any members of my own species could be so crazy. Then I begin to think a bit harder. Why is it that they and I should respond to what is virtually the same environment in such a radically different way? Even if I could see any purpose in such a ridiculous pursuit, nothing would induce me to hazard life and limb in undertaking it at such a high risk. Of course not, because, though the environment may have changed little, and though I may at birth have possessed a potential for development very similar to theirs, the world I have created in my own mind is vastly different.

I can just imagine the Easter Island children of a few hundred years ago learning, just as I did, by play, by make-believe, by listening to the words of presumably infallible grown-ups, and above all by exploration, about the world in which they were to spend the rest of their lives. A major difference between us is that not only their world but also the world from which their parents and grandparents had distilled their accumulated wisdom was to all intents and purposes about fifteen miles long and about half as wide. I can see practically the whole of it from this crater. I can even detect with the naked eye, and very clearly with binoculars, a number of *moai*, those huge stone statues for which the island is best known (Figure 55). I can see another extinct volcano at Rano Raraku, perhaps ten miles away, in the flank of which is situated the quarry where these huge monuments were carved out of the volcanic rock, at least six hundred of them, weighing anything up to eighty tons apiece.

It's inconceivable that such children, brought up among these vast effigies with recognisable human attributes yet so much larger and more forbidding than their own parents, could have regarded them with anything but awe. Whatever their function, or whoever

FIGURE 54. RANO KAU, EASTER ISLAND. View to the south-west from the 'Birdman Village' of Orongo on the lip of the volcanic crater. Note the 'birdman' carvings in bas-relief in the foreground and the islands of Motu kao-kao, Motu-iti and Motu-nui. Photo by Mark Appleton, 1986.

FIGURE 55. 'MOAI' AT AHU NAUNAU, ANAKENA, EASTER ISLAND.
Thor Heyerdahl's excavations in progress, January, 1986. Photo by the author.

they were supposed to represent, any model of the world which failed to accord them not only recognition and respect but even reverence would have no chance of making any sense.

Indeed, if the perennial argument about the influence of heredity or environment, 'nature' or 'nurture', the inborn or the learned, were really an 'either-or' argument, one would surely be forced to come down on the side of culture as the main determinant of the differences between me and those generations of islanders whose mental models of the world must have been so unlike my own. My understanding of what I'm looking at from the rim of this crater really is of a different order from theirs. I know, as they could not know, that, if I were to travel a hundred and fifty times as far as I can now see from my vantage-point to the eastern tip of the island, I should still not have reached the South American mainland. I can quantify what they could only guess at in the vaguest way. Yes, I am different! The meanings and values I attach to what I see are different, and this is principally because I've been subjected to a richness of information-input which far surpasses that of the best-informed of those islanders of long ago.

But of course heredity and environment provide not alternative but complementary sources of explanation for our behaviour, and I mustn't allow my evident superiority of comprehension to blind me to the similarities which are to be found in our respective habits of thought. If those islanders were persuaded of the existence of supernatural forces which could influence their daily lives, what about that divine power which I had been brought up to believe resided in that medieval building not thirty yards from my own childhood home? If they could embrace the fallacy of attributing the life force to anthropoid shapes carved out of the volcanic rock, what about those inanimate objects to which I also had attributed qualities of personality, even if only poetically, like mountains and railway engines? If they allowed themselves to be guided by emotion rather than by reason, how many times had I followed in their footsteps? If I'm honest with myself I'm bound to confess that a very important component of my own environmental perception is non-rational. I *feel* my habitat as much as I think it.

And if the comparison between myself and these long-departed islanders has to be seen in terms of similarities as well as of differences, how does the retired professor, standing on the rim of this crater, see his relationship with that little boy who began, some sixty-five years earlier, to explore his habitat in the shrubberies of an English rectory garden, in the long grass of a country churchyard and in the hedges and ditches which bounded the surrounding fields? Intellectually we are two very different people, though we answer to the same name, but I am well aware of many feelings which have proved remarkably durable. However sophisticated my rational interpretation of my habitat may have become, my emotional response to the sight of a newt rising with outstretched fingers and toes from the mysterious depths of a mirky pond, or of white clouds casually brushing the tips of lombardy poplars, or of green trailing weeds sweeping from side to side in a clear river and exposing alternating patches of its gravelly bed hasn't really changed. And when I think of the places with which I associate these visual experiences I begin to understand more clearly what Eyles means when he writes about relating things to 'the totality of people's lives' (Eyles, 1985, 6).

All those images which meant so much to me in childhood are still as evocative of that feeling of which I remember being so deeply aware when I used to watch the sunset from a dormitory window in the Glebe House and which I later came to recognise as what other people called 'beauty' for want of a better word. The intellectual, rational, interpretative dimension has since been added, bringing with it a new satisfaction, but not to the exclusion of the old. Above all, the adventure of exploring is still charged with that simple, primitive and intensively pleasurable excitement which I can now identify as the driving-force motivating, indeed impelling me to inform myself about the world around me. It's just as well that that has survived, because without it our species would surely have perished long, long ago.

WORKS REFERRED TO IN THE TEXT

Anderson, Charles R., *The Magic Circle of Walden* (New York: Holt, Rinehart & Winston, 1968).

Appleton, J. H. (Jay) (1960) 'The Communications of Watford Gap, Northamptonshire', *Institute of British Geographers, Transactions and Papers*, 28:215-24.

The Geography of Communications in Great Britain (Oxford: O.U.P. for the University of Hull, 1962).

Disused Railways in the Countryside of England and Wales: Report to the Countryside Commission (London: HMSO, 1970).

The Experience of Landscape (London & New York: Wiley, 1975. Paperback, Hull University Press 1986).

The Poetry of Habitat (Hull: Department of Geography, University of Hull, 1978).

The Symbolism of Habitat: an Interpretation of Landscape in the Arts (Seattle: University of Washington Press, 1990).

Awdrey, The Rev. W., *Thomas, The Tank Engine* (Leicester: Edmund Ward, n.d.).

Baker, R. Robin, *Human Navigation and the Sixth Sense* (London: Hodder & Stoughton, 1981).

Balling, J.D. and Falk, J.H., 'Development of visual preference for natural environments' (*Environment and Behavior*, 14, 5-28, 1982).

Bannerman, Helen, *The Story of Little Black Sambo* (London: Grant Richards, 1899).

The Story of Little Black Mingo (London: Nisbet, 1901).

Blythe, Ronald, *Akenfield:Portrait of an English Village* (London: Allen Lane/Penguin, 1969).

Bourassa, Steven C., *The Aesthetics of Landscape* (London and New York: Belhaven Press, 1991).

Burgess, Jacquelin. See Harrison, Carolyn et al.

Burke, Edmund, *A Philosophical Enquiry into the Origins of our Ideas on the Sublime and the Beautiful* (1757) (Bicentenary edition by Boulton J. T., London: Routledge & Kegan Paul (1958)).

Butler, Samuel, *Erewhon: or Over the Range* (London & Edinburgh: Trübner, 1872).

Campbell, H. J.,*The Pleasure Areas* (London: Eyre Methuen, 1973).

Carroll, Lewis, *Alice in Wonderland* (London: Macmillan, 1865).

Cervantes, M., *The Adventures of Don Quixote*, with illustrations by Harry G. Theaker (London & Melbourne: Ward Lock & Co., 1929 edition).

Cook, Helen and Bill, *Khaki Parish: Our Love - Our War* (Worthing: Churchman Publishing, 1988) Paperback, Hodder & Stoughton (1989).

Crabbe, George, 'A Lover's Journey' in *Tales* (London: Hatchard, 1812).

Daniels, S. and Cosgrove, D., *The Iconography of Landscape* (Cambridge: Cambridge University Press, 1988).

Darwin, Charles R., *The Voyage of The Beagle* (London: Henry Colborn, 1839) (Everyman's Library Edition, London: 1955).

On the Origin of Species by Means of Natural Selection, or the Preservation of Favoured Races in the Struggle for Life (London: Murray, 1859).

Dewey, John, *Experience and Nature* (London: Allen & Unwin, 1929).

Dryden, John, *The Works of Virgil:* containing his Pastorals, Georgics, and Aeneis (London: Tonson, 1697).

Eyles, John, *Senses of Place* (Warrington: Silverbrook Press, 1985).

Fitzgerald, Edward, *The Rubáiyát of Omar Khayyám* (London: Quaritch, 1859).

Frey, S. R., 'A Series of Gardens' (*Landscape Architecture*, 76:54-61, 1986).

Gear, Jane, *Perception and the Evolution of Style: a New Model of Mind* (London: Routledge, 1989).

Gibson, J.J., *The Ecological Approach to Visual Perception* (Boston: Houghton Mifflin, 1979).

Goodman, Nelson, *Ways of Worldmaking* (Hassocks: The Harvester Press, 1978).

Gray, Thomas, *Elegy Written in a Country Churchyard* (London: Dodsley, 1751).

Harrison, Carolyn; Limb, Melanie; and Burgess, Jacquelin, 'Recreation 2000, Views of the Country from the City' (*Landscape research* 11, 1986, 19-24).

Hildebrand, Grant, *The Wright Space* (Seattle: University of Washington Press, 1991).

Housman, A.E., *A Shropshire Lad* (London: Kegan Paul, 1896).

Howard, Peter, *Landscapes: The Artists' Vision* (London: Routledge, 1991).

Hudson, Brian, 'The View from the Verandah: Prospect Refuge and Leisure' (*Australian Geographical Studies* 31(1):70-78, 1993).

Hudson, Robert, *Inside Outside Broadcasts* (Newmarket: R & W Publications, 1993).

Humphrey, Nicholas, 'The Illusion of Beauty' (Perception, 2:429-39, 1972).

Kaplan, Rachel and Kaplan, Stephen, *The Experience of Nature* (Cambridge: Cambridge University Press, 1989).

Kaplan, Stephen and Kaplan, Rachel, *Cognition and Environment: Functioning in an Uncertain World* (New York: Praeger, 1982).

Kipling, Rudyard, *Just So Stories for Little Children*, illustrated by the author (London: Macmillan, 1902).

Landscape Santa Fe, New Mexico (later Berkeley, California) (1951-)

Landscape Research South Kilworth: Landscape Research Group (1976-).

Limb, Melanie. See Harrison, Carolyn et al.

Lorenz, Konrad, *King Solomon's Ring* (London: Methuen 1952, 1964 edition consulted).

Lowenthal, David, *The Past is a Foreign Country* (Cambridge: C.U.P., 1985).

Lowenthal, D. and Prince, H.C., 'The English Landscape' (*Geographical Review*, 54, 309-46, 1964) and 'English Landscape Tastes' (*Geographical Review*, 55, 186-22, 1965).

Milne, A.A., *Winnie the Pooh*, with decorations by E.H. Shepard (London: Methuen, 1926).

The House at Pooh Corner with decorations by E.H. Shepard (1928).

Oldham, J.B., *A History of Shrewsbury School. 1552-1952* (Oxford: Blackwell, 1952).

Omar Khayyám. See Fitzgerald, Edward.

Orians, Gordon H., 'Habitat Selection: General Theory and Applications to Human Behavior' in Lockard, J. (Ed.), *Evolution of Human Social Behaviour* (New York: Elsevier, 1980).

Blackbirds of the Americas (Seattle: University of Washington Press, 1985).

'An ecological and evolutionary approach to landscape aesthetics' in Penning-Rowsell, E.C. and Lowenthal, David (Eds) (1986).

Paulson, Ronald, *Literary Landscapes: Turner and Constable* (New Haven and London: Yale U.P., 1982).

Penning-Rowsell, Edmund C. and Lowenthal, David (Eds), *Landscape Meanings and Values* (London: Allen & Unwin, 1986).

Prince, Hugh C. See Lowenthal and Prince.

Rölvaag, O.E., *Giants in the Earth* (New York: Harper & Brothers, 1929).

Ruskin, John, *The King of the Golden River, or the Black Brothers* illustrated in colour by A. H. Baxter. (London: Geographia, n.d.).

Virgil, *Georgics*. See Dryden, John.

INDEX